CHEMOINFORMATICS: THEORY, PRACTICE, & PRODUCTS

CHEMOINFORMATICS: THEORY, PRACTICE, & PRODUCTS

B. A. BUNIN
Collaborative Drug Discovery, San Mateo, CA, U.S.A.

B. SIESEL
Merrill Lynch & Co., San Francisco, CA, U.S.A.

G. A. MORALES
Telik Inc., Palo Alto, CA, U.S.A.

J. BAJORATH
Rheinische Friedrich-Wilhelms-Universität, Bonn, Germany

 Springer

A C.I.P. Catalogue record for this book is available from the Library of Congress.

ISBN-10 1-4020-5000-3 (HB)
ISBN-13 987-1-4020-5000-8 (HB)
ISBN-10 1-4020-5001-1 (e-book)
ISBN-13 987-1-4020-5001-5 (e-book)

Published by Springer,
P.O. Box 17, 3300 AA Dordrecht, The Netherlands.

www.springer.com

Printed on acid-free paper

TABLE OF CONTENTS

SUBJECT APPENDICES

**Drug Discovery Informatics Registration Systems
and Underlying Toolkits (Appendices 1 and 2)**

Content Databases (Appendices 3–7)

Drug, Molecule, and Protein Visualization (Appendices 8–10)

Modeling and Algorithms (Appendices 11–17)

FOREWORD

Chemoinformatics: Theory, Practice & Products covers the theory, commercially available packages and applications of Chemoinformatics. Chemoinformatics is broadly defined as the use of information technology to assist in the acquisition, analysis and management of data and information relating to chemical compounds and their properties. This includes molecular modeling, reactions, spectra and structure-activity relationships associated with chemicals. Computational scientists, chemists, and biologists all rely on the rapidly evolving field of Chemoinformatics. *Chemoinformatics: Theory, Practice & Products* is an essential handbook for determining the right Chemoinformatics method or technology to use. There has been an explosion of new Chemoinformatics tools and techniques. Each technique has its own utility, scope, and limitations, as well as meeting resistance to use by experimentalists. The purpose of *Chemoinformatics: Theory, Practice & Products* is to provide computational scientists, medicinal chemists and biologists with unique practical information and the underlying theories relating to modern Chemoinformatics and related drug discovery informatics technologies.

The book also provides a summary of currently available, state-of-the-art, commercial Chemoinformatics products, with a specific focus on databases, toolkits, and modeling technologies designed for drug discovery. It will be broadly useful as a reference text for experimentalists wishing to rapidly navigate the expanding field, as well as the more expert computational scientists wishing to stay up to date.

It is primarily intended for applied researchers from the chemical and pharmaceutical industry, academic investigators, and graduate students.

The purpose of "Chemoinformatics: Theory, Practice, & Products" is to provide scientists with practical information and a fundamental understanding of the latest chemoinformatics technologies applied to drug discovery and other applications. Given an ever-expanding list of drug discovery informatics tools available to the modern researcher, understanding the underlying theories, organizing and summarizing the tools for best practices should be broadly useful. It is intended to be a regularly used text.

Chemoinformatics is broadly defined as information associated with molecules: both theoretical and experimental. This ranges from molecular modeling to reactions to spectra to structure-activity relationships associated with molecules. Chemoinformatics has the potential to revolutionize synthesis, drug discovery, or any science where one wants to optimize molecular properties. Computational scientists, chemists, and biologists all rely on the rapidly evolving field of chemoinformatics. The terms chemoinformatics and cheminformatics are often used interchangeably. As of July, 2006, the term "Cheminformatics" is leading "Chemoinformatics" ~306,000 to ~164,000 in a Google search (thanks to Phil McHale for his original suggestion). Despite this difference in Google popularity, we use chemoinformatics throughout this book because cheminformatics is frequently mis-interpreted as an abbreviation of the expression "Chemical informatics". As we will discuss, "Chemical informatics" has originally been used in a different context (and it is also not a very meaningful term).

Chemoinformatics, which can be viewed as either a subset or superset of Drug Discovery Informatics, has emerged as an interdisciplinary field of science of importance to chemists and biologists as well as computational scientists. Computational scientists use chemoinformatics tools to design and refine better models. Medicinal chemists use chemoinformatics tools to design and synthesize better compounds. Biologists use chemoinformatics to prioritize compounds for screening and assays for development. The drug discovery process is often analogized to the tale of the three blind men and the elephant where each "sees" a different beast by grabbing the tail, trunk, or side. The appropriate development of new and use of existing chemoinformatics tools is often directly a function of a specific problem … and problem solver. Thus having a centrally-compiled resource describing relevant chemoinformatics tools allows researcher to find the appropriately shaped "hammer" for their "nail."

"Chemoinformatics: Theory, Practice, & Products" provides the basic toolkits. It is a handbook that one can consult to determine the chemoinformatics method or technology of choice to use. The book covers the theory behind the methodologies *as well* as the practical information on commercially available products. The goal is to provide the perspective of computational chemists in a format accessible to experimentalists, too. Thus, there are sections on the underlying theory as well as sections overviewing the modern commercially available software and applications to provide the information of interest to computational scientists as well as to the broader audience of experimentalists.

There has been an explosion of new chemoinformatics tools and techniques. Each technique has some utility, scope, and limitations, as well as resistance to use by experimentalists. There is no compilation describing all the modern tools that are available. This guide will allow both experts and non-experts to know how and when to best use these technologies.

"Chemoinformatics: Theory, Practice, & Products" is intended for chemists, biologists, and computational scientists. It is basically for anyone interested in chemoinformatics for either synthesis or drug discovery. This includes the individuals at the companies mentioned in the book who work in the field of chemoinformatics (MDL, Accelrys, Tripos, CambridgeSoft, etc.) as well as the computational chemistry or drug design departments at biotechnology and pharmaceutical companies engaged in small molecule drug discovery and those using chemoinformatics for materials discovery too.

The book can be useful as a reference book for the experienced chemoinformatics expert or as a text to introduce the new student to the field. The information from the leading commercial suppliers is covered and organized into tables to help a wider range of scientists benefit from the revolution in informatics technologies in their day-to-day work. It is a reference of what is known as well as a guidebook to define what is possible with modern chemoinformatics technologies.

A quick disclaimer. Although a range of areas were covered including over a hundred product and methods, it is not possible to include everything under the sun. A more specialized book could be written entirely about any one of the seventeen subject appendices. Obviously tradeoffs had to be made between scope and depth of coverage. Furthermore, although it is inevitable that products and technologies will evolve over time, many of the most useful products are now mainstays of the modern chemoinformatics arsenal such as CAS-Scifinder, Beilstein, ChemDraw, Marvin, smiles strings, and Lipinski calculations – just to name a few. In addition to these well known products, there are often alternative products available with different specifications which are also described herein. Thus even as new trends emerge, the general state of modern chemoinformatics (and drug discovery informatics) is fundamentally represented. It is interesting to see the range of products that have historically been available as well as the evolution of new product areas such as gene-family wide SAR databases, data-pipelining, and metabolism predictors, just to name a few.

Perhaps most notable of the new initiatives is the publicly funded PubChem effort. A road map of existing products is useful both to differentiate new products and to prioritize the most important areas to focus future innovation. Understanding the landscape of existing products should be particularly useful to the buyers and sellers of chemoinformatics and drug discovery informatics technologies. Where might the field go in the future? With the emergence of open source software products in the broader software marketplace (for products like Linux, Apache, and MySQL), the integration of community-based tools with commercial tools has been a recently increasing phenomena. Similarly, the increasing number of openly available databases and tools emerging from the publicly funded initiatives such as the human genome project provide a fertile frontier for future innovation that combines the best of community and commercial chemoinformatics tools in new ways.

1. CHEMOINFORMATICS THEORY

The theoretical part of this book is intended to provide a general introduction into this still young and rapidly evolving scientific discipline. In addition, it is meant to provide a basis for researchers interested in applying products and tools that are detailed in the later sections. Therefore, it is attempted to outline some of the most relevant scientific concepts on which current chemoinformatics tools are based and provide some guidance as to which methodologies can be applied in a meaningful way to tackle specific problems. As such the theoretical sections are first and foremost written for practitioners with various scientific backgrounds and also students trying to access chemoinformatics tools. Therefore, the description of mathematical formalisms will be limited to the extent required to achieve a general understanding. In addition, rather than trying to provide an extensive bibliography covering this field, it is attempted to limit citations to key publications and contributions that are accessible to a readership with diverse scientific backgrounds.

As a still evolving discipline, chemoinformatics is an equally interesting playground for method development, chemical and drug discovery applications, and interdisciplinary research. This makes this field a rather exciting area to work in and it is hoped that the information provided herein might encourage many scientific minds to actively contribute to its further development.

1.1 CHEMOINFORMATICS – WHAT IS IT?

The term chemoinformatics (which is synonymously used with cheminformatics) was introduced in the literature by Brown in 1998 and defined as the combination of "all the information resources that a scientist needs to optimize the properties of a ligand to become a drug" (Brown 1998). Following this definition, both decision support by computer and drug discovery relevance are crucial aspects. On the other hand, the term chemical informatics was already used much earlier and generally understood as the application of information technology to chemistry, thus lacking a specific drug discovery focus. In addition, the chemometrics field focuses on the application of statistical methods to chemical data in order to derive predictive models or descriptors. Although these definitions and areas of research still co-exist, it appears to be increasingly difficult to distinguish between them, in particular, as far as method development is concerned. Therefore, it has recently been suggested to more broadly define chemoinformatics and include the types of

1

TABLE 1.1.   The spectrum of chemoinformatics

Chemical data collection, analysis, and management
Data representation and communication
Database design and organization
Chemical structure and property prediction (including drug-likeness)
Molecular similarity and diversity analysis
Compound or library design and optimization
Database mining
Compound classification and selection
Qualitative and quantitative structure-activity or – property relationships
Information theory applied to chemical problems
Statistical models and descriptors in chemistry
Prediction of *in vivo* compound characteristics

---

computational methodologies and infrastructures in the chemoinformatics spectrum that are shown in Table 1.1 (Bajorath, 2004).

This extended definition does no longer imply that chemoinformatics is necessarily linked to drug discovery and takes into account that this field is still evolving. Moreover, approaches that are long established as disciplines in their own right are also part of the chemoinformatics spectrum. This is well in accord with other views that chemoinformatics might largely be a new rationalization of tasks in chemical research that have already existed for considerable time (Hann and Green 1999). In fact, chemoinformatics research and development should be capable of adopting established scientific concepts and putting them into a novel context. Given the above topics, good examples for this might include, among others, the use of quantitative structure-activity relationship (QSAR) models for computational screening of large compound databases or the use of fragments of active compounds (so-called substructures) as a starting point for the design of targeted combinatorial libraries. In its extended definition, chemoinformatics includes all concepts and methods designed to interface theoretical and experimental programs involving small molecules. This is a crucial aspect because there is little doubt that the evolution of chemoinformatics as an independent discipline will much depend on its ability to demonstrate a measurable impact on experimental chemistry programs, regardless of whether these are in pharmaceutical research or elsewhere.

1.2   CHEMO- VERSUS BIO-INFORMATICS

There is little doubt that data explosion in chemistry and biology has been the major driver for the development of chemoinformatics and bioinformatics as disciplines. In the 1990s the advent of high-throughput technologies in biology (DNA sequencing) and chemistry (combinatorial synthesis) had caused much of the need for efficient computational infrastructures for data processing, management, and mining. In biology raw DNA sequences were the primary data source, whereas in chemistry rapidly

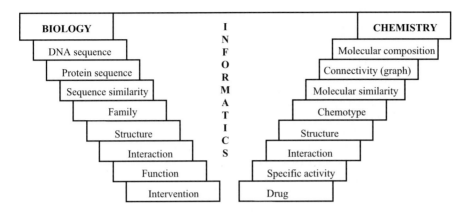

Figure 1.1.   Hierarchy of bio- and chemoinformatics research

growing compound databases produced by combinatorial synthesis techniques provided previously unobserved amounts of primary data (structures) and secondary data (screening results). Over a relatively short period of time, however, both bio- and chemo-informatics have developed well beyond data processing and management and have become research-intense disciplines.

How distinct are bio- and chemoinformatics as disciplines? Figure 1.1 summarizes the topics at different stages of typical bio- or chemoinformatic analysis.

Clearly, proceeding from molecular composition to two- and three-dimensional structure and function or activity presents a number of similar challenges, regardless of whether the starting point is a DNA sequence or the chemical element distribution of a molecule. From an algorithmic point of view, many tasks for this type of analysis in biology and chemistry are often much more similar than one might think, considering the diversity of biological and chemical applications. Thus, many algorithms and computational techniques used in chemoinformatics, as will be described herein, are also used for many applications in bioinformatics. For example, cluster algorithms are not only applied to classify compound databases but also to analyze expression data sets. Similarly, statistical algorithms are used to correlate compound structures with specific activities and also to correlate expression patterns and experimental conditions in microarray analysis. Thus, "similar algorithms – diverse applications" is a general theme in applied informatics research. Such insights are also consistent with recent trends in the life science area where bio- and chemoinformatics are beginning to merge. This is particular relevant for drug discovery where chemical and biological information needs to be integrated as much as possible to be ultimately successful and where the boundaries between different disciplines have become rather fluid. For example, it could hardly be decided whether the development of relational databases that link compound structure with assay data, biological targets, and pharmacological information would be a bio- or chemoinformatics project. Thus, informatics research and development in the life sciences is expected to become much more global in the future.

1.3 SCIENTIFIC ORIGINS

Given the above outline of the chemoinformatics field, one should also review the scientific roots that have laid the foundation for the development of chemoinformatics as a research discipline, beyond data management. In the 1960s, efforts begun to correlate compound structures and activities in quantitative terms by modeling linear relationships with the aid of molecular descriptors (Hansch and Fujita 1964; Free and Wilson 1964). These studies provided the basis of quantitative structure-activity relationship (QSAR) analysis, which was ultimately extended to multi-dimensional QSAR in 1980 (Hopfinger 1980). Also in the 1960s, chemical structures were first stored as computer files in searchable form by Chemical Abstract Services, thus providing a basis for structure retrieval and searching (Willett 1987). During the 1970s, methods for two-dimensional substructure (Cramer et al. 1974) and three-dimensional pharmacophore searching (Gund 1977) were developed, which made it possible to search compound databases for desired structural motifs or active molecules. In the 1980s, clustering methods were adapted for chemical applications, became very popular for the classification of molecular data sets, and were applied to explore similarities from various points of view (Willett 1988). The concept of molecular similarity itself became a major research topic in the late 1980s (Johnson and Maggiora 1990). Molecular similarity analysis extended conventional QSAR approaches where the influence of small compound modifications on activity is studied. Thus, relationships between molecular structure (and properties) and biological activity were beginning to be explored from a more global point of view. During the 1990s, the concepts of molecular diversity and dissimilarity complemented similarity analysis and algorithms were developed for the design of chemically diverse compound libraries (Martin et al. 1995) and selection of diverse compounds from databases (Lajiness 1997; for a compendium of interesting personal accounts of the early days of molecular similarity and diversity research, see Martin 2001). Although many other efforts have – without doubt – significantly contributed to and helped to shape chemoinformatics, as we understand it today, it is evident that two major themes have largely dominated the development of this discipline: chemical data organization and mining and, in addition, the exploration of structure-activity relationships (from many different points of view).

1.4 FUNDAMENTAL CONCEPTS

1.4.1 Molecular descriptors and chemical spaces.   The majority of chemoinformatics methods depend on the generation of chemical reference spaces into which molecular data sets are projected and where analysis or design is carried out. The definition of chemical spaces critically depends on the use of computational descriptors of molecular structure, physical or chemical properties, or pharmacophores. Essentially, any comparison of molecular characteristics that goes beyond simple structural comparison requires the calculation of property values and the application

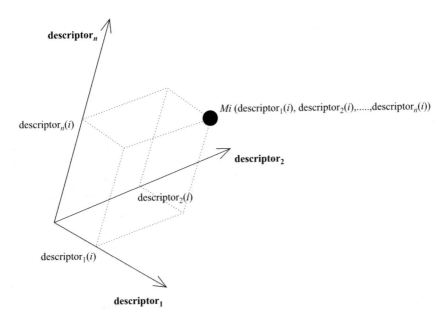

Figure 1.2.   *N*-dimensional chemical space with a molecule *M* at position *i*

of mathematical models. In chemical space design, each chosen descriptor adds a dimension to the reference space, as illustrated in Figure 1.2.

For each molecule, calculation of *n* descriptor values produces an *N*-dimensional coordinate vector in descriptor space that determines its position:

Molecule i: $M(i) = \sum_1^j d_j(i)$

Hundreds if not thousands of molecular descriptors have been designed for chemical applications (for an encyclopedic descriptor compendium, see [Todeschini and Consonni 2000]) that can be divided into different types or classes. Some examples are given in Table 1.2. Descriptors are frequently divided into 1D, 2D, or 3D descriptors, dependent on the dimensionality of the molecular representation from which they can be calculated, as illustrated in Figure 1.3.

The design and complexity of different types of descriptors often varies dramatically. Among very simple descriptors are, for example, 2D structural fragments that have, however, high predictive value in many applications because they implicitly account for diverse molecular properties (such as complexity, polarity, hydrophic character etc.). Topological indices, for example, go beyond simple structural fragment description and introduce a next level of abstraction. To give an example, for a molecule containing

TABLE 1.2.   Different types of molecular descriptors

Descriptor category	Examples
Physical properties	Molecular weight
	logP(o/w)
Atom and bond counts	Number of nitrogen atoms
	Number of aromatic atoms
	Number of rotatable bonds
Pharmacophore features	Number of hydrogen bond acceptors
	Sum of van der Waal surface areas of basic atoms
Charge descriptors	Total positive partial charge
	Dipole moment from partial charges
Connectivity and shape descriptors	Kier and Hall molecular shape indices
Surface area and volume	Solvent-accessible surface area

1D

$C_{22}H_{24}ClFN_4O_3$ $\longrightarrow$ Number of carbon atoms

2D

Number of rotatable bonds
log P(o/W)
Molecular connectivity index

3D

Solvent–accessible surface area
Van der Waals volume

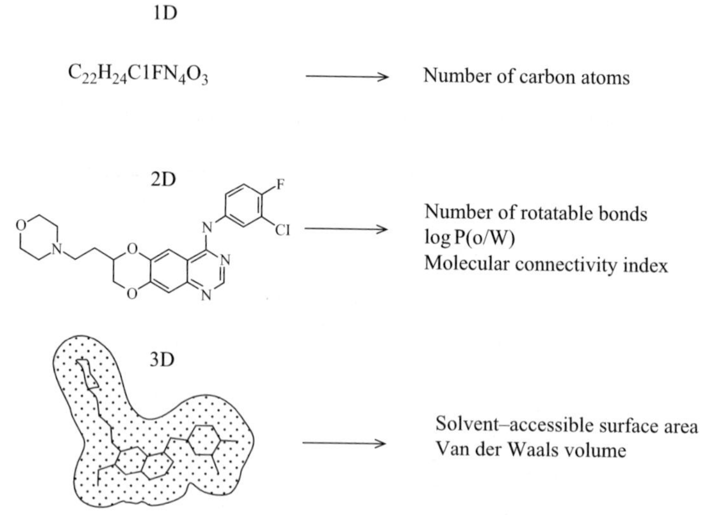

Figure 1.3.   Examples of descriptors classified according to dimensionality
(adapted from Bajorath 2002)

n atoms and m bonds, so-called first and second order kappa shape indices (Kier 1997) are calculated as follows:

$$\kappa_1 = \frac{n(n-1)^2}{m^2}$$

$$\kappa_2 = \frac{(n-1)(n-2)^2}{m^2},$$

These indices are also 2D descriptors because they require a molecular drawing (or graph) in order to determine the number of bonds.

Other descriptor designs can become increasingly complex. Contributions of different types of descriptors can also be combined into composite formulations, for example, descriptors combining molecular surface and charge information such as charged partial surface area (CPSA) descriptors (Stanton and Jurs 1990). As an example, two of these descriptors, $PNSA_1$ and $PPSA_1$, capture the sum of the solvent-accessible surface area (SAA) of all negatively and positively charged atoms in a molecule, respectively:

$$PNSA_1 = \sum_{n-} SAA^-$$

$$PPSA_1 = \sum_{n+} SAA^+$$

Thus, calculation of these descriptor values for a molecule involves the separate calculation of atomic charges and SAAs.

1.4.2 Chemical spaces and molecular similarity. There are no generally preferred descriptor spaces for chemoinformatics applications and it is usually required to generate reference spaces for specific applications on a case-by-case basis, either intuitively, based on experience, or by applying machine learning techniques to automate and optimize descriptor selection for a given problem. However, descriptors are ultimately selected for chemical space design, n descriptors always produce an n-dimensional reference space, as discussed above, into which compound sets can be mapped. In meaningful chemical space representations, similar compounds should map to similar regions, in other words, their intermolecular distance should be small. This represents a basic interpretation of the similarity concept. Table 1.3 lists examples of conventional distance functions that are used for these calculations.

Here n_i and n_j are the number of descriptor values for molecules i and j, respectively, and n_{ij} is the number of common values. D_{ij} is the distance between molecules i and j, D the average distance, and n the total number of molecules.

It should be noted that the general understanding of molecular similarity goes beyond simple structural similarity and extends to biological activity, in accord with the so-called Similar Property Principle (Johnson and Maggiora 1990) postulating that molecules having similar structures and properties should also exhibit similar activity

TABLE 1.3. Distance functions

Hamming distance	$HD = \sum_{i=1}^{n} x_i \oplus y_i,$
Euclidean distance	$ED = \sqrt{\sum_{i=1}^{n} (x_i - y_i)^2}$
Average distance	$AD = \dfrac{\sum_{i=1}^{n} \sum_{j=1}^{n} D_{ij}}{n(n-1)}$

In the formula of the Hamming distance, $\oplus$ means "exclusive disjunction" and detects non-identical values. In the formula of the average distance, D_{ij} is the distance between molecules i, j and n the total number of molecules.

TABLE 1.4.   Similarity coefficients

Tanimoto coefficient	$Tc = n_{ij}/(n_i + n_j - n_{ij})$
Dice coefficient	$Dc = 2n_{ij}/(n_i + n_j)$
Cosine coefficient	$Cc = n_{ij}/(n_i n_j)^{1/2}$

(which is often – but not always – true). Thus, molecules that are located closely together in chemical reference space are often considered to be functionally related, which is one of the hallmarks of molecular similarity analysis.

If descriptor combinations are expressed as bit strings (often called fingerprints, as described in more detail later on), each test molecule is assigned a characteristic bit pattern, and pair-wise molecular similarity can be assessed by quantifying the overlap of bit strings using various similarity metrics (coefficients). Examples are shown in Table 1.4.

In these formulations, n_i and n_j are the number of bits set on for molecules i and j, respectively, and n_{ij} is the number of bits in common to both molecules.

The values of these similarity coefficients range from zero (i.e., no overlap; no similarity) to one (i.e., complete overlap; identical or very similar molecules). In chemoinformatics, the most widely used metric is the Tanimoto coefficient.

1.4.3 Molecular similarity, dissimilarity, and diversity. How are similarity and diversity related to each other? As discussed, similar molecules can be identified by application of distance functions and analysis of nearest neighbors in chemical space. Diversity analysis, on the other hand, attempts to either select different compounds from a given population or, alternatively, evenly populate a given chemical space with candidate molecules. This can also be accomplished using distance functions by only selecting compounds that are at least a pre-defined minimum distance away from others or – in diversity design – by trying to maximize average inter-compound distances.

An alternative approach to diversity selection and design is to divide the descriptor axes into evenly spaced value intervals, a process called "binning", which produces n-dimensional subsections of chemical space (also called "cells", as discussed in a later section). Then it can be monitored how these cells are populated with compounds that are projected into chemical space. In diversity selection, one would attempt, for example, to select a representative compound from each populated cell; in diversity design, one would try to populate cells as evenly as possible with computed molecules. As will be discussed in the next section, such segment- or cell-based design strategies can, in practial terms, only be applied to low-dimensional descriptor spaces; otherwise, the vast majority of cells would remain empty, thereby preventing a meaningful analysis.

Molecular diversity is a global concept, which is applicable to the analysis of large compound distributions, but not to the study of pair-wise molecular relationships. This is in contrast to molecular similarity analysis, which explores pair-wise relationships, the exploration of which is more local in nature. For example, one tries to find compounds similar to a given reference molecule or study the compound population within a limited region of chemical space. From this point of view, the inverse of molecular similarity is not diversity, but rather "dissimilarity", which is local in nature (addressing the question which molecule in a collection is most dissimilar from a given compound or set of compounds). Like similarity, dissimilarity calculations can focus on the exploration of pair-wise compound relationships (e.g., distances in chemical space). When similarity metrics are applied, the dissimilarity d between two molecules i and j is thus defined as, for example:

$$d_{ij} = 1 - Cc(i,j) \quad \text{or} \quad 1 - Tc(i,j)$$

Dissimilarity analysis plays a major role in compound selection. Typical tasks include the selection of a maximally dissimilar subset of compounds from a large set or the identification of compounds that are dissimilar to an existing collection. Such issues have played a major role in compound acquisition in the pharmaceutical industry. A typical task would be to select a subset of k maximally dissimilar compounds from a data set containing n molecules. This represents a non-trivial challenge because of the combinatorial problem involved in exploring all possible subsets. Therefore, other dissimilarity-based selection algorithms have been developed (Lajiness 1997). The basic idea of such approaches is to initially select a seed compound (either randomly or, better, based on dissimilarity to others), then calculate dissimilarity between the seed compound and all others and select the most dissimilar one. In the next step, the database compound most dissimilar to these two compounds is selected and added to the subset, and the process is repeated until a subset of desired size is obtained.

1.4.4 Modification and simplification of chemical spaces. High-dimensional chemistry spaces might often be too complex for carrying out meaningful and interpretable analyses. One reason for this is that major areas or subsections of high-dimensional chemical space might not be populated with compounds and thus remain "empty". Another reason is that correlation effects between selected descriptors dramatically distort the reference space, which often (but not always) complicates the analysis of compound distributions. Therefore, it is generally attempted to either design low-dimensional reference spaces, simplify high-dimensional spaces, or reduce their dimensionality. Descriptor correlation is a very common effect. For example, the number of carbon atoms in a molecule (a very simple 1D descriptor) correlates with molecular weight, hydrophobicity etc. In fact, it is rather difficult to find a set of completely uncorrelated descriptors. Compound analysis or design in low-dimensional spaces has the added bonus that it is often possible to further reduce the dimensionality to three without too much loss of information so that one can visualize the results. Visualization of chemical space representations, even if only approximate, is in general of high value, as it permits a more intuitive analysis of molecule distributions and makes it possible to complement computations with chemical knowledge and experience. There are several different ways to simplify chemical spaces or produce low-dimensional representations, as discussed in the following.

Regardless of space dimensionality, it is generally important to scale selected descriptors because their value ranges may substantially differ for a given data set. Descriptors with large value ranges will dominate those having smaller ones and distort the analysis (i.e., a very "long" coordinate axis in chemical space might render "short" axes nearly "invisible"). Therefore, auto-scaling or variance scaling with mean centering is typically applied:

$$d_i' = \frac{(d_i - d_{av})}{\sigma}$$

Here d_i is the descriptor value of molecule i, d_{av} the average (or mean) value of the entire data set, the σ standard deviation, and d_i' the scaled value of descriptor d for molecule i. This procedure ensures that all chosen descriptors have similar value ranges (i.e., that descriptor axes have comparable length) and thus prevents space distortions.

The most common way to generate low-dimensional reference spaces is dimension reduction of original descriptor spaces, as illustrated in Figure 1.4. This process attempts to define a low-dimensional representation that captures data variability to the same or a similar extent as the original descriptor space.

Dimension reduction relies on the assumption that high-dimensional descriptor spaces have at least some intrinsic redundancy, which is in most cases true as a consequence of descriptor correlation effects. There are two major categories of methods to facilitate dimension reduction, for which different algorithms are available. One class of methods attempts to identify those descriptors that are most important for representing the original data set to use them and the relationships they form between objects for lower-dimensional representations. An example for this approach is multi-dimensional scaling (Agrafiotis *et al.* 2001). The other type of methods attempts to generate new descriptors for low-dimensional spaces by combining important contributions from the original ones. A representative method is principal component analysis that processes descriptor variance and co-variance matrices of compound sets and

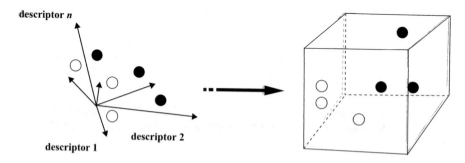

Figure 1.4. Dimension reduction. The figure illustrates the transformation of an *n*-dimensional descriptor space into an orthogonal three-dimensional space formed by three non-correlated descriptors either selected from the original ones or derived from them as new composite descriptors. Test compounds are shown as white or black dots

ultimately calculates novel composite descriptors as linear combinations of the original ones (Xue and Bajorath 2000):

$$PC1 = c_{1.1}d1 + c_{1.2}d2 + c_{1.3}d3 + \cdots + c_{1.n}dn + \text{const}_1$$
$$PC2 = c_{2.1}d1 + c_{2.2}d2 + c_{2.3}d3 + \cdots + c_{2.n}dn + \text{const}_2$$
$$\vdots$$
$$PCn = c_{n.1}d1 + c_{n.2}d2 + c_{n.3}d3 + \cdots + c_{n.n}dn + \text{const}_n$$

Here the coefficients c reflect the importance of each descriptor (within each component) to capture data variance. Principal component analysis removes descriptor correlation effects and the resulting components account for data variance in descending order (i.e., the first accounts for more than the second, the second more than the third, and so on). A possible result would be that for an original 20-descriptor space, the first five or six principal components account for greater than 95% of the variance within the molecular data set, thus permitting the generation of an orthogonal reference space reduced to five or six dimensions.

An alternative to dimension reduction is the use of composite and uncorrelated descriptors that are suitable for the design of information-rich yet low-dimensional chemical spaces. An elegant example is presented by the popular BCUT (Burden-CAS-University of Texas) descriptors (Pearlman and Smith 1998). BCUTs are a set of uncorrelated descriptors that combine information about molecular connectivity, inter-molecular distances, and other molecular properties. BCUT spaces used for many applications are typically only six-dimensional and can frequently be further reduced to 3D representations for visualization purposes by identifying those BCUT axes around which most compounds map.

Simplification of n-dimensional descriptor spaces is another alternative to dimension reduction. This can be accomplished, for example, by analysis of descriptor value distributions in large databases. In statistics, the median of a value distribution is defined as the value that separates it into two equal halves (above and below the median). Thus, a descriptor with continuous database value range can be transformed into a binary scheme where a test compound with a descriptor value above or equal to the median is assigned a transformed descriptor value of "1" and a compound with a descriptor value below the median a transformed value of "0" (Godden et al. 2004). Binary descriptor transformation retains the dimensionality of the original space but greatly simplifies it because the length of each descriptor axis is either zero or unity. Thus, this simplified descriptor space is also scaled. As further discussed in the following, both low-dimensional and binary-transformed descriptor spaces have been proven very useful for partitioning analyses and compound classification or design.

1.5 COMPOUND CLASSIFICATION AND SELECTION

The classification of molecular data sets according to pre-defined criteria is one of the central themes in chemoinformatics. Methods designed to classify molecules are applied in database organization and mining and also provide a basis for selection of compounds according to diversity, property, or biological activity criteria.

1.5.1   Cluster analysis.   As mentioned before, clustering has been one of the roots of the chemoinformatics field and continues to be widely applied. Clustering methods are often divided into non-hierarchical and hierarchical techniques and hierarchical methods are further divided into divisive or agglomerative clustering. Hierarchical-divisive methods start from a large cluster containing all compounds ("top-down"), whereas hierarchical-agglomerative techniques begin from single-tons ("bottom-up"). Hierarchical clustering builds relationships between clusters in subsequent steps, which means that the composition of each cluster depends on the one from which it was derived. Non-hierarchical clustering methods organize compounds into an initially defined number of independent clusters, which is often accomplished by calculating nearest neighbor distributions in chemical space. Molecules can be expressed as descriptor vectors and for each cluster, a center vector can be calculated (e.g., as an average position) that distinguishes different clusters from each other. New molecules are assigned to clusters based on their distances from different cluster centers in descriptor space. Figure 1.5 illustrates these different clustering approaches.

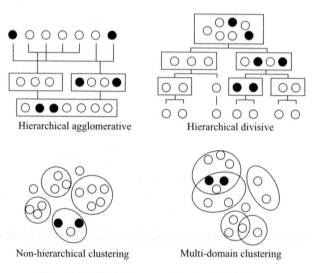

Hierarchical agglomerative   Hierarchical divisive

Non-hierarchical clustering   Multi-domain clustering

Figure 1.5.   Clustering approaches described in the text
(adapted from Kitchen *et al.* 2004)

Fuzzy clustering methods that have recently become popular are distinct from traditional clustering techniques in that molecules are permitted to belong to multiple clusters or have fractional membership in all clusters. A potential advantage of such classification schemes is that more than one similarity relationship can be established by cluster analysis.

Hierarchical clustering also depends on the chosen "linkage scheme" that determines the way inter-cluster distances are calculated. For example, based on "single linkage", inter-cluster distance is defined as the minimum distance between members of two clusters. By contrast, "complete linkage" calculates the maximum distance between members in two different clusters. Furthermore, for all clustering methods, clustering levels and cluster occupancy present additional variables. For example, too many clustering steps will result in sparsely populated clusters and too few in densely populated ones, both of which will distort molecular similarity relationships derived from clustering (which means that molecules within the same cluster should be "similar"). Therefore, level selection algorithms are typically applied in order to determine calculation parameters that balance clustering levels and cluster occupancy.

Among non-hierarchical methods, Jarvis-Patrick clustering (Jarvis and Patrick 1973) has been popular early on in chemical database analysis. It is a nearest neighbor method: two molecules are included in the same cluster if they share a pre-defined minimum number of nearest neighbors. However, the method has been found to produce rather unevenly sized clusters, often too large or too small. Another popular non-hierarchical method is k-means clustering where k clusters are randomly seeded, cluster averages or means are calculated in descriptor space, and molecules are re-assigned to other clusters if their position is closer to those means than to the one of their initial cluster. This clustering technique is fast but depends on the initial random seeding of clusters with test compounds and the choice of k. Over the years, agglomerative-hierarchical methods, in particular, Ward's clustering (Ward 1963), have become more popular in chemistry because this approach has been shown to produce more balanced cluster levels and distributions than non-hierarchical methods, resulting in more reliable classification of similar molecules.

As discussed, clustering algorithms generally involve distance comparisons between compounds or between compounds and cluster centers in chemical space, which renders calculations increasingly demanding as the compound databases grow in size. Given currently available computational power, classifications methods that involve exhaustive pair-wise compound or distance comparisons can be applied to thousands of compounds but become prohibitive when databases further increase in size by orders of magnitude.

1.5.2 Partitioning. In contrast to clustering techniques, partitioning algorithms do not rely on pair-wise molecular and distance comparison and can therefore be applied to very large compound source databases. Rather than comparing molecular positions, partitioning methods establish a coordinate or reference system in chemical space that ultimately defines the position of each compound based on its calculated descriptor coordinates. Compounds that populate the same partitions or sub-sections of chemical space are considered similar. Partitioning in low-dimensional descriptors spaces, generated either by use of BCUT descriptors or dimension reduction techniques, has become a very popular approach. Cells are generated by dividing orthogonal (uncorrelated) descriptor axes into regularly spaced intervals or bins, as illustrated in Figure 1.6.

Regardless of whether clustering or partitioning algorithms are applied, compound classification calculations are often carried out to provide a basis for compound selection from large data sets. Major strategies include diversity- or activity-oriented selection, as illustrated in Figure 1.7. Diversity-based selection aims at generating a small representative subset of a compound collection. In this case, it

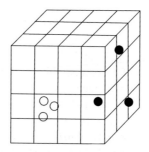

Figure 1.6.   Axes of a low-dimensional orthogonal chemical space are binned in order to produce cells for partitioning. White dots represent similar compounds

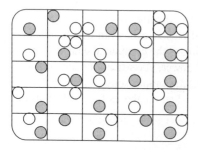
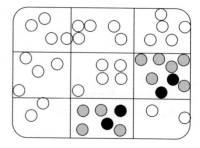

Figure 1.7.   Diversity-based (left) and activity-based (right) compound selection from partitions. White dots represent database compounds, gray dots selected database compounds, and black dots known active compounds

is attempted to generate evenly populated partitions or clusters from which representative compounds are selected in order to mirror the overall diversity distribution in chemical space. By contrast, in activity-based selection, known active compounds are added to the source database prior to clustering or partitioning. Database compounds mapping close to known actives are then selected as candidates for testing to identify new hits.

Cell-based partitioning can not only be used for compound selection but also to aid in combinatorial diversity design. In this case, a chemical descriptor space is defined and "empty" partitions are generated by binning. Test compounds are then enumerated on the computer based on reaction schemes and selected to evenly populate these partitions.

In addition to cell-based partitioning, statistical partitioning methods are widely used for compound classification. One of the most popular approaches is recursive partitioning (Rusinko *et al.* 1999), a decision tree method, as illustrated in Figure 1.8. Recursive partitioning divides data sets along decision trees formed by sequences of molecular descriptors. At each node of the tree, a descriptor-based decision is made and the molecular data set is subdivided. For example, a chosen descriptor could simply detect the presence or absence of a structural fragment in a molecule. Alternatively, the

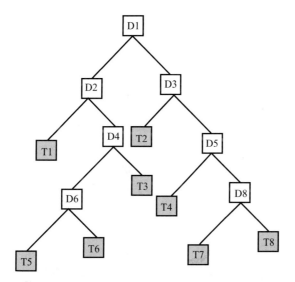

Figure 1.8.   Decision tree. Shown is a rudimentary tree structure (D, descriptors; T, terminal nodes) for recursive partitioning. Terminal nodes are shaded gray.

compounds could be divided according to molecular weight (e.g., equal or greater than 400 or less). Very many different descriptors can be utilized and decision tree methods are computationally very efficient and applicable to very large data sets. Typically, a learning set would consist of active and inactive compounds and one would search for descriptor combinations that enrich active compounds in certain terminal nodes. The so derived descriptor pathways can then be used to search compound databases for other active compounds. Therefore, statistical partitioning methods such as recursive partitioning are also very attractive tools for the analysis of HTS data sets and the extraction of descriptor-activity relationships from them that can serve as predictive models of specific biological activities.

1.5.3 Support vector machines In addition to more "traditional" classification methods like clustering or partitioning, other computational approaches have recently also become popular in chemoinformatics and support vector machines (SVMs) (Warmuth *et al*. 2003) are discussed here as an example. Typically, SVMs are applied as classifiers for binary property predictions, for example, to distinguish active from inactive compounds. Initially, a set of descriptors is selected and training set molecules are represented as vectors based on their calculated descriptor values. Then linear combinations of training set vectors are calculated to construct a hyperplane in descriptor space that best separates active and inactive compounds, as illustrated in Figure 1.9.

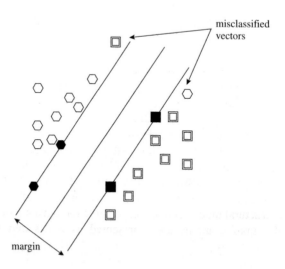

Figure 1.9.   SVM-based hyperplane. Two classes of molecules are separated in descriptor space by a hyperplane ($H(x) = 0$) with margins ($H(x) = \pm 1$). Support vectors are shown as filled objects and used to construct the hyperplane and define its margins

A polynomial function is applied to return the inner products of descriptor vectors and the separating hyperplane is defined as

$$H(x) = 0 = \langle w, x \rangle + b$$

Here b is the distance between the hyperplane and the origin and w the distance between vector x and the hyperplane. The margin of the hyperplane is defined as its minimum distance to any training set vector. A small number of vectors constituting the margin are called support vectors and are sufficient to construct a hyperplane that separates the remaining data points into two subsets. SVM calculations and hyperplane construction can initially be carried out using small training sets and additional data can be added in a step-wise manner to further refine the prediction scheme.

1.6   SIMILARITY SEARCHING

Searching for compounds in databases that are similar to query molecules is one of the most widely applied molecular similarity-based approaches. Commonly used similarity search tools have different levels of complexity, as discussed in the following.

1.6.1   Structural queries and graphs.   A simple but very popular form of similarity searching is the detection of structural fragments or substructures that are shared by query and database compounds. Figure 1.10 illustrates the idea of substructure searching.

In medicinal chemistry, substructure searches are often carried out to find analogs of active compounds in databases. Contemporary substructure search methods are mostly based on dictionaries or look-up tables of molecular fragments or fragment-type descriptors. Substructure search queries based on dictionaries of predefined molecular fragments can be transformed into an easily machine-readable format such as the Simplified Molecular Input Line Entry specification widely known as SMILES code (Weininger 1988). As illustrated in Figure 1.11, SMILES encodes 2D representations of molecules as linear strings of special alpha-numeric characters for atoms, their chemical character, bonding patterns, branch points, stereo centers etc. These query strings can be easily compared to those of database compounds.

Substructure searching has also been coupled with statistical analysis to identify molecular fragments that are associated with biological activity of test compounds (Roberts *et al.* 2000). For example, substructures can be taken from series of active compounds and their frequency of occurrence in compound databases can be determined. Then the statistical significance of this frequency of occurrence in active molecules and database compounds can be compared, which might indicate whether or not a specific structural motif could be responsible for a given biological activity.

In molecular graphs, atoms are represented as nodes and bonds as edges. Conventional 2D representations of compound structures are typical graphs, but graph representations of molecular connectivity can also be much simpler than that (as described below). Common substructures can also be determined by systematic mapping of corresponding node positions in graphs, which is called the analysis of "subgraph isomorphism". However, computationally this is a much more expensive

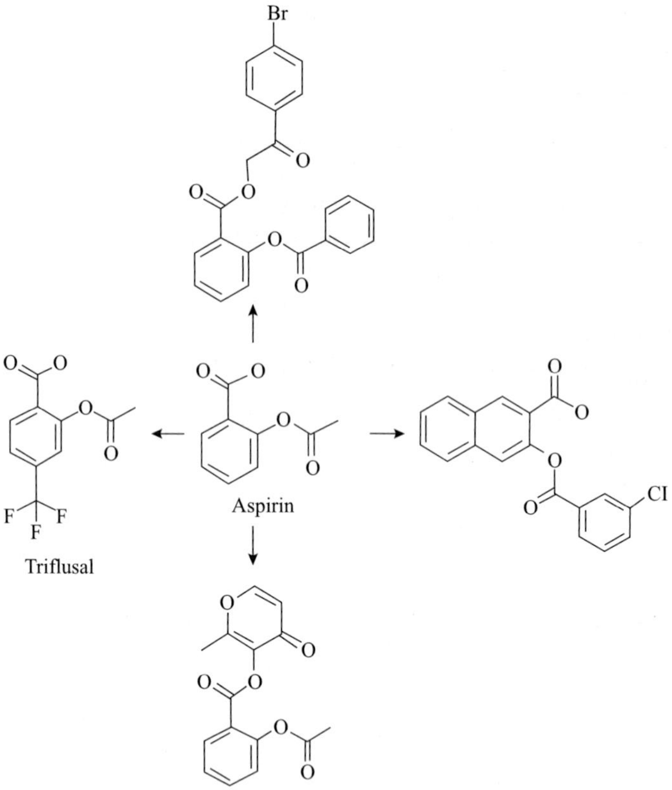

Figure 1.10.   Example of compounds containing Aspirin as a substructure that
can be used as a query for database searching

procedure than dictionary-based (or "grammatical") approaches to substructure
analysis. Therefore, similarity searching using molecular graphs has generally been
computationally prohibitive, until recently when reduced graphs were developed for
these purposes (Gillett *et al*. 2003). In reduced graphs, nodes do not represent atoms
but features such as functionally important groups or whole ring systems, which
reduces the level of detail of the representations of intra-molecular connectivity, as
illustrated in Figure 1.12. Thus, such simplified graph representations become more
suitable for node matching procedures and similarity searching.

1.6.2   Pharmacophores.   Going beyond 2D substructures, pharmacophores are
defined as spatial arrangements of atoms or groups that are responsible for biological
activity. Such geometric arrangements of important moieties or groups are often used as
3D queries for databases searching. Pharmacophores are most often derived from computed conformations or conformational ensembles of active compounds and less so from

Structures	Strings
	c1ccccc1
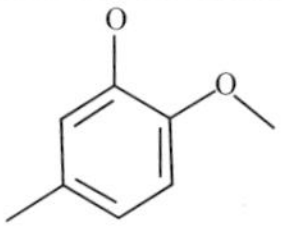	Oc1cc(C)ccc1OC
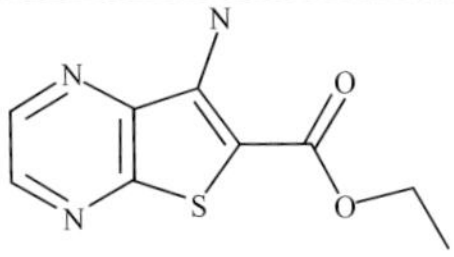	s1c2[nH0]cc[nH0]c2c(N)c1C(=0)OCC
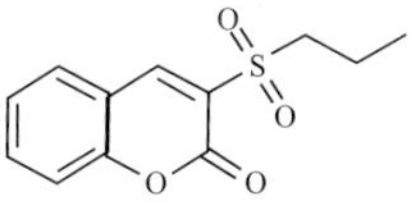	[S+2]([O–])([O–])(CCC)C1=Cc2ccccc2OC1=O
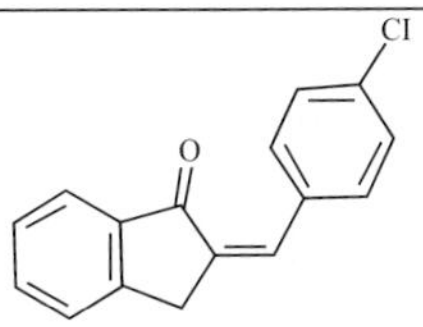	C1c1ccc(cc1)C=C1Cc2ccccc2C1=O
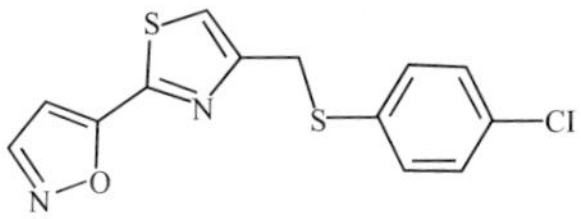	Clc1ccc(SCc2[nH0]c(sc2)c2o[nH0]cc2)cc1
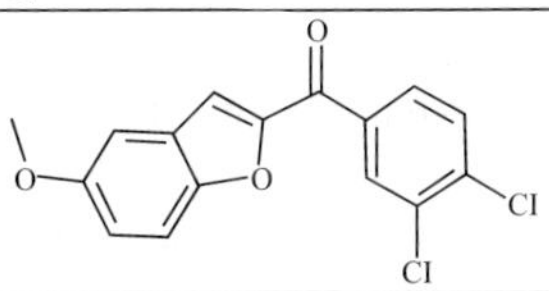	Clc1ccc(cc1C1)C(=O)c1oc2ccc(OC)cc2c1
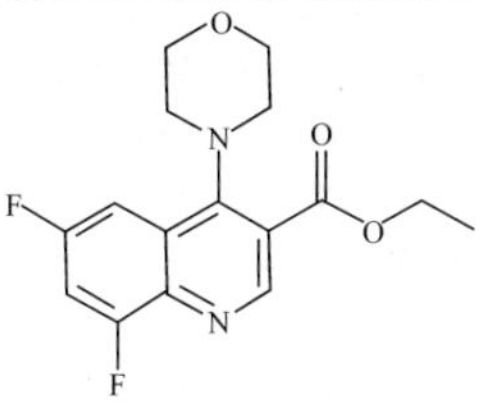	Fc1cc(F)c2[nH0]cc(c(N3CCOCC3)c2c1)C(=O)OCC

Figure 1.11.   Examples of structures and corresponding SMILES strings

CHEMOINFORMATICS

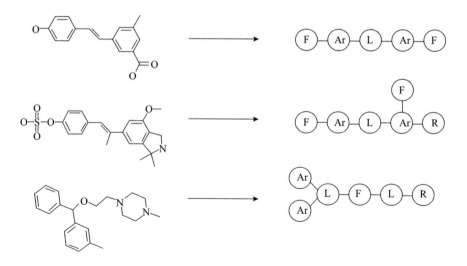

Figure 1.12.   Examples of reduced graphs. Nodes corresponding to aromatic rings (Ar), aliphatic rings (R), functional groups (F) and linking groups (L) are shown (adapted from Gillet *et al.* 2003)

experimentally determined structures of ligands. Consequently, the majority of pharmacophore models used to identify similar compounds represent hypotheses of 3D features crucial for biological activity.

For database searching, pharmacophores are best defined by all possible distances between chosen groups or features (pharmacophore points). Therefore, as illustrated in Figure 1.13, they are best represented as a molecular graph (similar to reduced graphs). In this case, different from conventional graphs, however, nodes correspond to points (or centroids) and edges to inter-point distances, rather than bonds.

Graphs of query and test molecules can be compared by graph matching (subgraph detection) algorithms or systematic comparison of inter-feature distances. Two molecules are considered similar if their pharmacophores match for at least one predicted conformation. In order to explore conformational space and generate conformational ensembles, multiple compound conformations are typically generated by systematic conformational search (in increments) around rotatable bonds.

Three-point pharmacophores have traditionally been used for many applications but have recently been more and more replaced by four-point pharmacophores (Mason *et al.* 1999), which increases the complexity of the search but also the resolution of the pharmacophore analysis. This is the case because the additional point increases the total number of inter-point distances from three for a three-point pharmacophore to six for a four-point pharmacophore. Pharmacophore searching is further refined by assigning alternative features to each point (e.g., hydrogen bond acceptors, donors, or charged groups) and ranges to inter-point distances (rather than an exact distance). For example, five different features (e.g., atom types or groups) may be permitted for each point

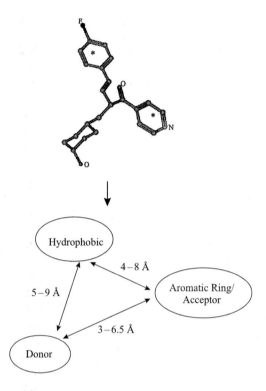

Figure 1.13.   Example of a possible 3-point pharmacophore representation

and ten inter-point distance intervals for each range, which makes it possible to capture thousands (if not millions) of similar yet distinct conformation-dependent pharmacophore arrangements.

1.6.3   Fingerprints.   Molecular fingerprints are widely used similarity search tools. They consist of various descriptors that are encoded as bit strings. As illustrated in Figure 1.14, bit strings of query and database compounds are calculated and quantitatively compared using similarity metrics such as the Tanimoto coefficient (Table 1.2). Fingerprint overlap between test compounds is regarded as a measure of molecular similarity. Thus, if the chosen coefficient reaches a pre-defined threshold value, compared molecules are considered to be similar.

In many fingerprint designs, each bit position accounts for a specific feature (for example, a structural fragment) and the bit is set on, if this feature is present in the molecule. Furthermore, value ranges of other molecular descriptors (e.g., molecular weight or the number of hydrogen bond acceptors) can also be incrementally encoded as bit strings. So designed fingerprints may consist of hundreds to thousands of bit positions. It is important to note that string representations of molecular

CHEMOINFORMATICS

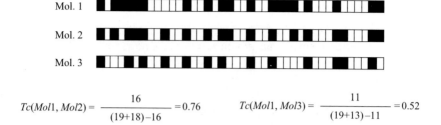

Mol. 1

Mol. 2

Mol. 3

$$Tc(Mol1, Mol2) = \frac{16}{(19+18)-16} = 0.76 \qquad Tc(Mol1, Mol3) = \frac{11}{(19+13)-11} = 0.52$$

Figure 1.14.   Model fingerprints and Tc comparisons

structure and properties also operate in chemical reference spaces, very similar to other compound classification and similarity-based methods, as discussed above. For example, a fingerprint capturing n descriptors using n bits represents a vector in n-dimensional descriptor space where the length of each dimension is either zero or unity (thus, analogous to binary transformed reference spaces, as discussed above). In other fingerprint designs, descriptor bit patterns may be mapped to overlapping bit segments, a process called folding or hashing. Consequently, each bit position can no longer be firmly associated with a specific feature making it difficult to interpret bit patterns in terms of descriptor settings.

Regardless of design specifics, fingerprints are often classified as 2D or 3D according to the dimensionality of the descriptors they encode. According to this scheme, fingerprints capturing structural fragments are considered 2D fingerprints. By contrast, pharmacophore fingerprints are 3D fingerprints. In these fingerprints, each bit position accounts for a specific multiple-point pharmacophore (Mason et al. 1999) and for a test molecule, large numbers of potential pharmacophores are encoded (many of which may not be related to biological activity). For pharmacophore fingerprinting, pharmacophores are also calculated based on systematic conformational search and it is assumed that increasing overlap of pharmacophore patterns in fingerprints corresponds to increasing molecular similarity.

For fingerprints and other similarity-based approaches, it is often discussed in the literature whether 2D or 3D methods (or descriptors) are generally superior. Intuitively, one might expect that 3D methods should be more realistic and perform better. However, in practice, this is not the case. Often, difficulties in accurately predicting active conformations of compounds and conformation-dependent 3D properties compromise the accuracy of 3D approaches. Accordingly, better results are frequently obtained with simpler yet more robust 2D methods. In fact, in many cases, 2D methods (for example, those making use of structural fragment-type descriptors) are surprisingly powerful, probably because they implicitly capture much chemical information. At present, it is not possible to answer the question whether 2D or 3D approaches perform overall better. In general, their relative performance does not only depend on the design of algorithms but also on the specifics of applications (e.g., molecular targets, compound classes etc) (Sheridan and Kearsley 2002).

1.7 MACHINE LEARNING METHODS

Computational methods involving machine learning techniques to gradually improve solutions to a given problem play an important role in chemoinformatics. For example, it is usually difficult, if not impossible, to predict or guess which types of descriptors (or descriptor combinations) are most suitable for a given search, classification, or compound design problem. Therefore, machine learning techniques are often used to facilitate descriptor selection, which can be done by systematically exploring various descriptor combinations and further refining those that give best intermediate results. Furthermore, machine learning methods are also applied to generate complex predictive models by iterative processing of molecular learning sets and subsequent application to compound databases (for example, to search for biologically active compounds or distinguish drug(-like) molecules from others). Two prominent machine learning techniques are discussed as examples here, genetic algorithms (GAs) and neural networks (NNs). Statistical methodologies such as recursive partitioning, as discussed above, are often also considered machine learning techniques, since they are capable of producing descriptor combinations and predictive models based on training sets that can then be applied to data mining.

1.7.1. Genetic algorithms. The design of GAs reflects basic principles of competitive biological evolution and population dynamics (Forrest 1993). Following these ideas, different parameters and model solutions to given problems are encoded in a chromosome and subjected to iterative random variations, thus generating a population. Solutions provided by these chromosomes are evaluated by a fitness function that assigns high scores to desired results. Chromosomes yielding best intermediate solutions are subjected to mutation and cross-over operations that correspond to random genetic mutations and gene recombination events, respectively. The resulting modified chromosomes represent the next generation and the process is continued until the obtained results meet a satisfactory convergence criterion of the fitness function. Figure 1.15 shows an example of a chromosome designed for descriptor and calculation parameter selection and outlines the GA optimization process.

In this example, descriptors and parameters for cell-based partitioning are optimized using principal component analysis as a dimension reduction technique. The chromosome was designed to consist of three bit segments, the first encoding 100 potential descriptors (one per bit position), the second encoding between one and 15 principal components for generation of low-dimensional descriptor space, and the third between one and 15 bins per principal component axis. During GA optimization, each chromosome-encoded combination of descriptors and calculation parameters of the initial population is submitted to principal component analysis of the descriptor space. Based on its calculated descriptor values and principal component coordinates, each molecule of a training set consisting, in this case, of different types of active compounds is projected into this space and cell population and occupancy rates are determined. For such calculations, a fitness function would typically favor input parameter combinations that

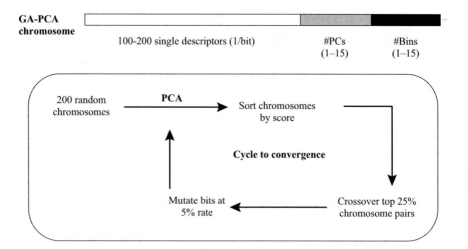

Figure 1.15.   GA-based principal component analysis (PCA) of descriptor combinations. The
chromosome encodes single descriptors and calculation parameters for PCA (#PCs, number of
principal components; #Bins, number of bins per component axis)

lead to co-partitioning of molecules sharing similar activity, while minimizing the
occurrence of mixed cells (i.e., cells containing molecules having different activities).
Combinations producing desired results receive high scores and the chromosomes
encoding these combinations are subjected to mutation and cross-over operations and
form the next generation. Once preferred descriptor combinations and calculation
parameters have been identified, these settings can be applied to other test sets or database mining. It is important to note that GAs usually produce reasonable to good solutions to a problem under investigation, but not necessarily an optimal one (if it exists).

1.7.2   Neural networks.   Artificial neural networks are computationally implemented as arrays of connected mathematical formulas or models that are organized
in different layers (Zupan and Gasteiger 1999). This type of architecture essentially
corresponds to arrangements of neurons and synapses in the nervous system.
Mathematical models are nodes and represent neurons and inter-node connections
(that can be weighted to assign different strengths to them) correspond to synapses.
NNs can simulate learning processes in different ways. In supervised learning, training sets are fed into NNs to distinguish different objects and their known properties
based on calculation of multiple variables or descriptors. Other objects with
unknown properties are then classified according to the obtained prediction schemes.
Supervised learning is usually achieved in what are called feed-forward NNs that
process data in several layers consisting of varying numbers of nodes. These nodes
are organized as input nodes, several layers of hidden nodes, and output nodes.
Figure 1.16 shows a schematic representation of the architecture of such feed-
forward NNs.

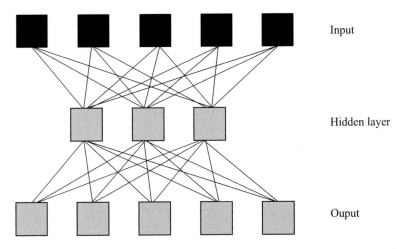

Input

Hidden layer

Ouput

Figure 1.16.   Schematic architecture of a three layer feed-forward network

Supervised learning using multi-layered NNs usually involves so called back-propagation of data in order to train the intermediate or hidden node layers for a given classification problem. Training is continued until a sufficiently accurate solution is obtained for the training set data, and the so derived node settings and connection weights are then used to classify new molecules. Typical tasks for supervised NN learning in chemoinformatics include, for example, distinguishing active from inactive compounds or drug-like molecules from non-drugs. A general and sometimes problematic feature of NN simulations is that the resulting classification models can usually not be interpreted or explained in physical or chemical terms (a situation often referred to as the "black box" character of NNs). On the other hand, a major advantage of NNs is their ability to capture and model non-linear relationships.

In contrast to supervised learning schemes, unsupervised learning does not depend on known properties of objects and a pre-defined classification scheme but attempts to identify relationships between objects and various computed features during the process of learning. In chemoinformatics, unsupervised learning is used, for example to identify molecular similarity relationships in compound data sets using various descriptors (without the need to exactly understand what features render molecules similar or not). Thus, unsupervised learning involves the de novo generation of classification models in the course of machine learning. Kohonen neural nets or self-organizing maps (SOMs) (Kohonen 1989) are an important example of unsupervised NN learning. SOMs consist of connected nodes that have a data vector associated with them. In molecular classification, descriptor vectors are calculated for test molecules and SOM nodes are assigned corresponding vectors, initially with random values. Then each test molecule is mapped to the node having the smallest distance to its descriptor vector in chemical space. During the learning phase, vectors of matching nodes and connected neighboring nodes are changed and made

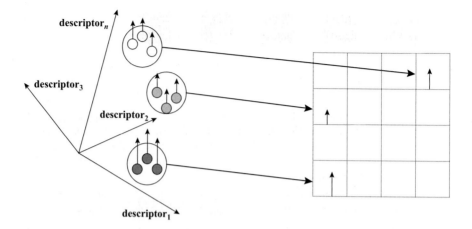

Figure 1.17.   SOM-projection of compounds from high-dimensional descriptor space onto a 2D array
of nodes. Groups of similar compounds map to similar nodes (and node vectors)

more similar to the one of the test molecule. This creates groups of similar nodes
that match test molecules having similar descriptor vectors. The learning process
continues by gradually reducing the connection weights and value
adjustments of neighboring nodes. These calculations generate larger numbers of
groups of similar nodes but reduce group size, which increases the resolution of the
molecular classification scheme. As illustrated in Figure 1.17, SOMs ultimately
assign similar molecules to regions of similar nodes and additional compounds can
be mapped based on their descriptor vectors. Thus, SOMs can be used as a clustering tool.

Since SOMs are capable of projecting compound distributions in high-dimensional
descriptor spaces on two-dimensional arrays of nodes, this methodology is also useful
as a dimension reduction technique, similar to others discussed above. SOM projec-
tions and the relationships they establish are usually non-linear, in contrast to, for
example, principal component analysis (that, as discussed, generates a smaller number
of new composite descriptors as linear combinations of the original ones).

1.8   LIBRARY DESIGN

The approaches discussed thus far have mostly been compound classification and
selection techniques. However, compound and library design strategies have played
an equally important role in chemoinformatics since the early 1990s, due to the addition of combinatorial chemistry to pharmaceutical R&D. For computational scientists, this brought along new challenges and a paradigm shift: rather than attempting
to optimize single hits or leads, by QSAR analysis or structure-based design techniques, and predict potent analogs, it became necessary to help with the design of
large compound libraries. Early on the concept of molecular diversity took center

stage and since then the compound library design field has further evolved and also experienced some significant changes, as discussed in the following.

1.8.1 Diverse libraries. Enumeration of combinatorial libraries refers to the process of systematically adding R-groups to molecular core structures, given pre-defined chemical reaction schemes (and reaction sites). Mimicking combinatorial reactions computationally requires the specification of points of diversification at reactive sites in core structures or scaffolds, the extractions of R-groups from available reagents (often called "clipping"), and the systematic combination of scaffolds and R-groups conforming to the reaction scheme. Figure 1.18 shows an example to illustrate this process of scaffold-based design.

It is evident that full enumeration of such libraries can quickly become prohibitive. For example, three points of diversity in a single scaffold with 100, 200, and 400 reagents/R-groups available for site 1, 2, and 3, respectively, would already results in $(100 \times 200 \times 400)$ or eight million possible combinatorial products. More complex reaction schemes and large reagent numbers can easily result in literally billions of potential products. Even full enumeration of moderately sized libraries is usually not meaningful because many potential products are likely to be very similar to each other and provide redundant chemical information. In drug discovery, the information content of large chemical libraries is a particularly critical issue because these libraries must be screened in order to identify novel active compounds, which can become a daunting and expensive task, despite the availability of high-throughput screening technology.

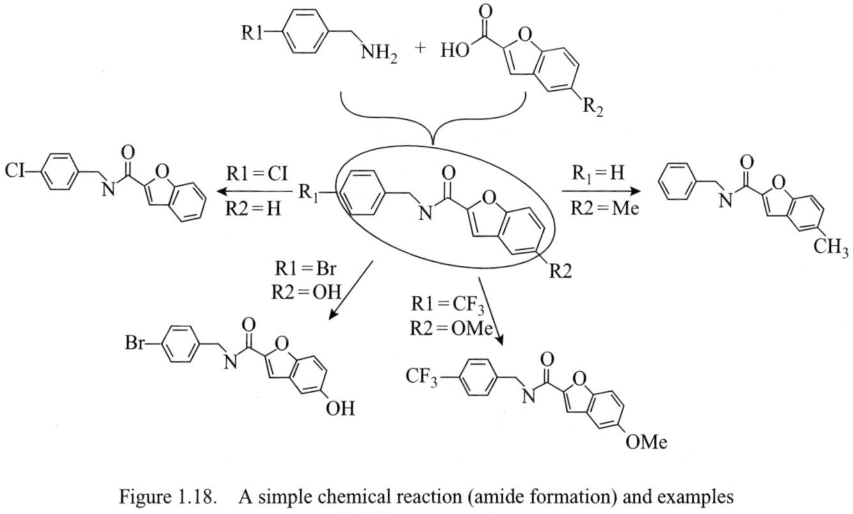

Figure 1.18. A simple chemical reaction (amide formation) and examples of scaffold/R-group combinations

Given this situation, it is not surprising that diversity design of combinatorial libraries has early on become a significant issue for chemoinformatics R&D and an intensely studied topic (Martin 2001). The major strategy of diversity design is to evenly sample however defined chemical space with compounds and build a library with synthetic candidates that is limited in size but represents the chemical diversity encoded in all possible products. In order to compute and sample diverse compounds, the same diversity metrics discussed above for compound selection are applied. Different strategies can be pursued. For example, cell-based algorithms can be applied to construct a descriptor space where cells are evenly populated with combinatorial compounds (which is assessed by monitoring cell population and occupancy rates). For these purposes, it is often critical to be able to visualize compound distributions in chemical space (which again explains the conceptual attractiveness of compound selection and design in low-dimensional chemical spaces). Another diversity design strategy is to first randomly sample some reaction products and then only accept others if they are sufficiently dissimilar, based on a distance criterion, from products that are already included in the library. However, this approach requires pair-wise distance comparisons, which ultimately limits library size. Regardless of the design strategy, an important point is that any diversity design depends on the property descriptors that are selected to generate the chemical reference space. Thus, there is also no absolute measure of diversity in library design, which – by necessity – includes at least some subjective elements. In addition to de novo design of combinatorial libraries, diversity design methods are also applied to complement or extend the diversity of existing compound collections by adding new molecules to be synthesized (Agrafiotis 2002), which often referred to as "hole filling". Methodologically, such efforts are closely related to dissimilarity analysis as applied in the context of compound selection and acquisition strategies.

1.8.2   Diversity estimation.   In library design, molecular diversity can be estimated in rather different ways (Gillet 2002). One possibility is a reactant-based diversity measure where the diversity of selected types of reagents is analyzed and diverse reagent subsets are selected for combinatorial exploration. This strategy is based on the premise that diversity in reagent spaces translates into combinatorial product space. Another possibility is to estimate overall diversity from the reaction products, which is computationally much more challenging, because there are many more possible products than reagents. These alternative strategies have been compared in a number of studies and a general conclusion is that product-based design – albeit computationally expensive – is more effective than reagent-based design in generating desired diversity distributions.

Product-based design can be pursued in different ways, taking into account the difficulties associated with library enumeration. For example, stochastic sampling of library products can be applied where statistically relevant subsets of a library are generated and analyzed to estimate the diversity distribution of the entire library (Agrafiotis 2002). So identified descriptor settings and design elements yielding a satisfactory degree of diversity can then be applied to generate the desired number of library products. Alternatively, a library design task can also be transformed into a compound selection exercise. In this case, a combinatorial library is virtually enumerated to the largest extent possible and diversity selection methods are applied to select a diverse or representative subset from the virtual library for synthesis. Once these compounds have been selected, reagents are extracted from them. This can be done by searching R-groups as substructures in a clipped reagent database or, more elegantly, by applying algorithms that are capable of retroactively dividing compounds into building blocks and reagents for specific chemical reactions (Lewell *et al.* 1998).

1.8.3 Multi-objective design.   Although diversity is a crucial aspect for library design (and the focal point of many studies), it is clearly not the only important design parameter. For example, the issue whether library compounds have desired chemical properties and are stable and non-toxic is equally important. In addition, compounds must be sufficiently soluble and should not exceed a certain molecular weight. Moreover, the availability and costs of reagents play an important role when planning library synthesis. Clearly, a number of objectives beyond diversity must be considered during the design phase and methods for simultaneous multi-objective library design have been introduced (Agrafiotis 2002; Gillet *et al.* 2002). For example, compound enumeration or selection can be monitored using a fitness function that evaluates a combination of different properties and weights them according to their assumed or desired relative importance:

$$F = f_1(P1) + f_2(P2) + f_3(P3) + \cdots + f_n(Pn)$$

The factors or weights *f* determine the relative importance of chosen properties *P* (e.g., diversity, solubility, availability, synthetic feasibility etc.). This type of fitness function could also be applied in machine learning using, for example, GAs. However, the weighted-sum-of-properties technique has limitations. Weights need to assigned, which can be rather subjective and, furthermore, additive fitness values or scores may not effectively distinguish between overall preferred and non-preferred solutions. Therefore, other GAs have also been developed where selected properties are improved independently (thus alleviating the need to subjectively assign factors) and where chromosomes are subjected to evaluation based on population dominance criteria (Gillet *et al.* 2002). Accordingly, individual property solutions within a population of chromosomes are ranked and those chromosomes that overall dominate are selected and further evolved. Ultimately, a population of overall equivalent multi-property solutions is obtained, each of which balances properties and design objectives in different ways, and from these alternative solutions, the most appropriate or convenient library design strategy can be selected.

1.8.4 Focused libraries.   A basic idea behind the design of diverse libraries has been that representative coverage of global chemically diversity would ensure that libraries could produce hits against many different therapeutic targets and thus be used as widely applicable screening tools. This principle is essentially valid, although chemical diversity encoded in large screening libraries still only represents small samples of theoretically accessible chemical space. However, over the years it has increasingly been recognized that smaller compound libraries designed for specific applications or "focused" on selected targets or compound activities are often much more effective discovery tools than diverse libraries. The general aim is to build libraries that are much smaller in size than diverse screening libraries but are significantly enriched with compounds having a high probability to exhibit specific activities. For example, while diverse library designs may typically contain between 100,000 to a million compounds, focused libraries may consist of only thousands of molecules (and sometimes even only hundreds). It is important to note that the generation of focused libraries, however designed, also involves chemical diversification. However, here global diversity analysis is substituted by exploring local diversity, for example, the chemical neighborhood of series of compounds already known to be active. On the other hand, focused library design can also employ molecular similarity analysis, for example, by selecting compounds from large screening collections that display a certain degree of similarity to structural motifs one attempts to focus on. In these cases, it must also be ensured that selected compounds are not too similar to each other (for example, by applying dissimilarity criteria). In general, limited or constrained diversity design is more suitable for the synthesis of new libraries, whereas similarity analysis better supports compound selection from larger pools. Figure 1.19 illustrates the relationship between diverse and focused compound distributions and chemical space coverage.

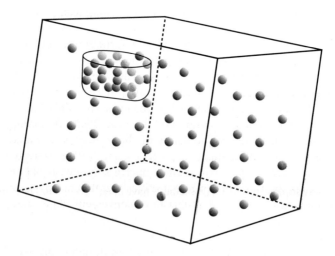

Figure 1.19.   Chemical space representation displaying diverse and focused
(boxed) compound sets

There are several (in part overlapping but distinct) approaches to the design of focused libraries (Schnur *et al*. 2004). An important distinction to be made is whether libraries should focus on a single target or a family of targets, which has become popular in recent years (for example, proteases, protein kinases, or G-protein-coupled receptors). Focusing on a target (or gene) family generally requires broader chemical coverage than a single target, but less specific information and fewer design constraints. First and foremost, known active compounds used as templates are chemically diversified and/or submitted to similarity searching. For example, using core structures of known protein kinase inhibitors as templates, combinatorial diversification produces a focused library that can be tested on this gene family with the hope that some derivatives will specifically interact with members for which effective and selective inhibitors have not yet been discovered. Suitable scaffolds for focusing on target families are often obtained by retrosynthetic combinatorial analysis (Lewell *et al*. 1998) of series of known actives, which may lead to the identification of so-called privileged substructures. These are structural motifs that frequently occur in active compounds and that provide attractive starting points for focused library design (Lewell *et al*. 1998; Mason *et al*. 1999).

For single targets, detailed pharmacophores (or pharmacophore fingerprints) are often used as chemical design constraints. Such pharmacophores can be extracted from known active compounds and also from 3D structures of target binding sites, if available. When focusing on target families, consensus pharmacophores can be derived by combining models obtained from studying different binding sites. Typically, the resolution of consensus pharmacophores is lower and design constraints are more relaxed, consistent with focusing on a target family. Another structure-based approach, which is applicable to single or multiple targets, first aims at finding database compounds that are overall consistent with the geometry and chemical features of binding site(s) (for example, by computational docking) and then uses these compounds as templates for focusing, as discussed above.

1.9 QUANTITATIVE STRUCTURE-ACTIVITY RELATIONSHIP ANALYSIS

The major goal of QSAR is the evaluation of molecular features that determine biological activity and the prediction of compound potency as a function of structural modifications. As such QSAR is generally limited to the study of closely related molecules or series of analogs. Since its introduction in the early 1960s the approach has been extensively used in medicinal chemistry to support lead optimization and has become one of the most popular computational methodologies. As discussed below, different and increasingly sophisticated QSAR methods have been developed. These methods do not necessarily require the study of biological activity but can target different types of physico-chemical features (e.g., absorption) as a function of molecular properties, giving rise to the field of QSPR (where P stands for property).

1.9.1 Model building. Classical QSAR analysis is based on two premises; first, that biological activity of compounds and differences in potency are a function of structural features and physico-chemical properties and, second, that the relationship between activity and molecular properties is linear. The first assumption is generally valid, although not all effects that are important for differential activity (e.g., compound solubility) are directly related to receptor ligand-interactions. The second premise clearly represents an approximation. Although many structure-activity relationships can indeed be well approximated using linear modeling, many others require the application of non-linear methods such as NN simulations. Accordingly, a number of current research activities focus on integrating QSAR and NN methods to permit both linear and non-linear modeling of structure-activity relationships (Tetko *et al*. 2001).

However, in the first instance, QSAR attempts to express compound potency as a linear function of various structural and property descriptors D with coefficients weighting their relative importance:

$$P = c_1 D1 + c_2 D2 + c_3 D3 + \cdots + c_n Dn + \text{const}$$

QSAR models are derived from learning sets consisting of series of compounds with different potency. In order to establish a linear relationship between combinations of descriptors and compound potency, the coefficients must be adjusted to fit the experimental data and linear regression is carried out (essentially minimizing the sum of differences between predicted and experimental values). In order to reduce the data range, one typically uses a negative logarithm of potency for modeling. Descriptors are often chosen based on knowledge (what features in test compounds are known to affect activity?) or intuition (what effects and corresponding descriptors typically determine receptor-ligand interactions?). However, machine learning methods and simulation techniques are also applied for descriptor selection. Regardless of descriptor selection methods, QSAR model building is an iterative process where various descriptor combinations are evaluated and correlated (redundant) descriptor contributions are removed. Determining the most appropriate number of descriptors (and their nature) is generally a non-trivial task. The choice of too few descriptors makes the model too general (with little, if any, predictive value), whereas too many descriptors render the model too specific for the training set (a process called over-fitting).

1.9.2   Model evaluation.   Before a QSAR model can be applied to predict the potency of new molecules, model evaluation and validation are essential. Initially, a squared correlation coefficient, r^2, is usually calculated to determine how well the actual linear function/model predicts the potency of training set compounds.

$$r^2 = 1 - \frac{\sum (p_{exp} - p_{calc})^2}{\sum (p_{exp} - p_{avg})^2}$$

where p_{exp} are experimental potency values, p_{avg} their average value, and p_{calc} the corresponding calculated ones. This coefficient ranges from zero (no fit) to one (perfect fit to data).

Once the quality of the linear model has been evaluated and confirmed, a more advanced validation step is carried out termed "leave-one-out cross-validation". Here, each data point (potency value) is removed once from the training set, the model is derived again for the remaining data points, and then applied to predict the value that was left out. The then obtained correlation coefficient is called cross-validated r^2 or q^2:

$$q^2 = 1 - \frac{\sum(p_{exp} - p_{pred})^2}{\sum(p_{exp} - p_{avg})^2}$$

The difference is that p_{calc} is replaced by p_{pred}, which are the predicted potency values that were not included in model derivation. Thus, while r^2 is a measure of the quality of the linear function to fit the training set data, q^2 is an indicator of the predictive power of the model. Like r^2, q^2 ranges from zero (no predictive value) to one (perfect predictions).

In 2D-QSAR, model building is based on 2D representation of molecules and 2D descriptors. However, it has become very common to generate 3D-QSAR models.

1.9.3   3D-QSAR.   Since compounds are active in three dimensions and their shape and surface properties are major determinants of their activity, the attractiveness of 3D-QSAR methods is intuitively clear. Here conformations of active molecules must be generated and their features captured by use of conformation-dependent descriptors. Despite its conceptual attractiveness, 3D-QSAR faces two major challenges. First, since bioactive conformations are in many cases not known from experiment, they must be predicted. This is often done by systematic conformational analysis and identification of preferred low energy conformations, which presents one of the major uncertainties in 3D-QSAR analysis. In fact, to date there is no computational method available to reliably and routinely predict bioactive molecular conformations. Thus, conformational analysis often only generates a crude approximation of active conformations. In order to at least partly compensate for these difficulties, information from active sites in target proteins is taken into account, if available (receptor-dependent QSAR). Second, once conformations are modeled, they must be correctly aligned in three dimensions, which is another major source of errors in the system set-up for 3D-QSAR studies.

A major milestone for the 3D- (and ultimately multi-dimensional) QSAR field has been the introduction of shape analysis in QSAR (Hopfinger 1980). Another milestone has been the development of Comparative Molecular Field Analysis or CoMFA (Cramer *et al.* 1988). Following this approach, a superposition of compounds on a three-dimensional grid is generated and conformation-dependent steric and electrostatic fields surrounding these compounds are calculated by probing their interactions with "pseudo-atoms" at grid points and mapping the resulting interaction energies to the grid. Figure 1.20 shows an example of such molecular field representations.

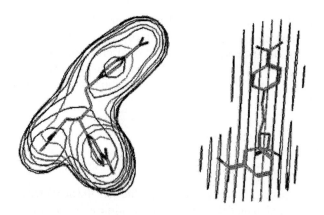

Figure 1.20.   Grid isosurface representation (orthogonal views) of the electrostatic
potential of a molecule

Based on the nature of the atom probes at each grid point and their interaction energies, a QSAR model is derived that expresses biological activity of test compounds as a function of probe-specific grid energies. Regions of similar grid interaction energies are plotted as contours. If the QSAR model suggests, for example, that increasing the charge of test compounds might improve potency, grid energy contours might suggest where a positively or negatively charged group could be placed in these compounds.

1.9.4   4D-QSAR.   The 4D-QSAR approach (Hopfinger *et al.* 1997) was designed to address the major problem of 3D-QSAR, the accurate prediction of bioactive compound conformations. This has been successfully done by calculating conformational ensembles of test compounds, rather than single conformations, which adds conformational space as a fourth dimension to QSAR. Although 4D-QSAR does not address the compound superposition or alignment problem, models can be optimized as a function of alternative alignments. In addition, the use of conformational ensembles often reduces the sensitivity to alignment errors, as long as these alignments are at least approximately correct. Even higher-dimensional QSAR models with additional degrees of freedom can be constructed, for example, by inclusion of binding site features.

4D-QSAR also utilizes grids for compound representation but not atom probes or interaction energies. Rather, compounds are divided into different regions based on the presence of a set of pre-defined features (that often occur in pharmacophores). Conformational ensembles are generated, for example, by molecular dynamics calculations. From these ensembles, special so-called grid cell occupancy descriptors are calculated. These descriptors are then correlated with compound activities by application of (partial least squares) regression analysis, and the most relevant grid cell occupancy descriptors are applied, usually in combination with other selected property descriptors, to establish the 4D-QSAR model.

1.9.5 Probabilistic methods. Other QSAR-like probabilistic approaches have also been developed for compound database mining. Binary QSAR (BQ) is discussed here as an example (Labute 1999). BQ is based on Bayes' theorem of conditional probabilities:

$$P(b|a) = \frac{P(ab)}{P(b)}$$

In this formulation, $P(b|a)$ is the probability that result "b" is obtained, if "a" has occurred, $P(ab)$ the probability that both a and b occur, and $P(b)$ the probability of result b alone. In BQ, Bayesian modeling is applied to learning sets consisting of active and inactive molecules to construct a probability function that estimates the probability of a molecule to be active, given its values of selected descriptors. Importantly, BQ uses a binary classification of activity where molecules are classified as either active or inactive given a pre-defined activity threshold value. Once successfully derived, the probability density function is applied to new molecules to estimate their relative probability (between zero and one) to be active. Thus, BQ is only semi-quantitative in nature. A major attraction of such probabilistic approaches is that they can be used to efficiently mine large compound databases for active compounds, similar to statistical partitioning algorithms. Due to its binary formulation of activity, BQ does not depend on the availability of highly accurate data sets for learning, which is another attractive aspect.

1.10 VIRTUAL SCREENING AND COMPOUND FILTERING

The term virtual screening (VS) is used to describe the process of screening large databases on the computer for molecules having desired properties and/or biological activity. In chemoinformatics, VS methods focus on small molecules, as opposed to computational docking of compounds to 3D structures of protein targets, and the term VS is sometimes synonymously used with (2D or 3D) similarity searching. However, any of the similarity-based compound classification or search methods discussed above can be adapted for VS (Bajorath 2002). There are several ways in which VS analysis can be applied.

1.10.1 Biologically active compounds. A major application of VS techniques is the identification of novel active molecules in large compound databases. Figure 1.21 illustrates how partitioning methods are adapted for this purpose.

These calculations generally start with a series of known active compounds that are added as search templates (or "bait" molecules) to a source database and compounds that are identified as similar to these templates based on VS calculations are selected as candidate molecules for experimental evaluation. Activity-oriented VS typically aims at identifying compounds that structurally differ from known templates but have similar activity. This is often done because known active compounds are difficult to develop, not easily chemically accessible (for examples, natural products), or already covered by other patents. In such cases, VS analysis attempts to identify molecular similarity relationships that balance structural and biological similarity in rather different ways, which is often a non-trivial task.

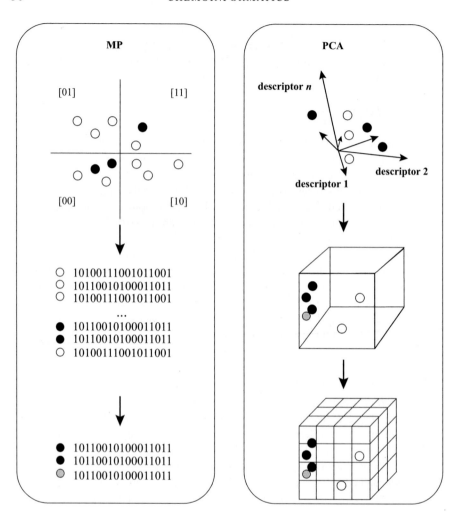

Figure 1.21. Application of a statistical partitioning algorithm (median partitioning, MP) and cell-based partitioning (based on PCA) to virtually screen compound databases. Known active compounds (or "baits") are shown in back and candidate molecules in gray (adapted from Stahura and Bajorath 2004)

1.10.2   Virtual and high-throughput screening.   As discussed above, VS is frequently used as a standalone database mining approach. However, this is not the only possibility. In fact, VS techniques can be successfully applied in concert with experimental high-throughput screening (HTS) in order to establish iterative or sequential screening schemes (Bajorath 2002). Here the major goal is to closely interface computational and experimental screening in order to reduce the number of compounds that need to be tested to identify novel hits. Figure 1.22 illustrates how iterative cycles of VS and HTS can be carried out to meet these goals.

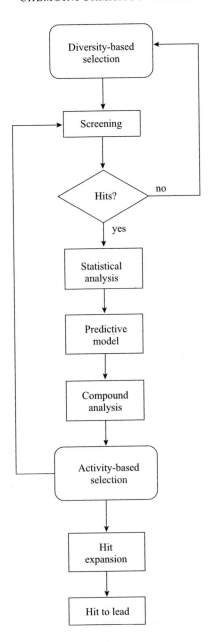

Figure 1.22.   Diversity and activity-based compound selection as part of integrated screening schemes
(adapted from Stahura and Bajorath 2004)

Following this scheme, a compound screening database is first reduced in size by computational selection of a representative subset that is screened experimentally to identify at least a few initial hits. These hits are then used as a starting point for VS to focus on activity-based selection of a limited number of compounds from a large screening collection that are then tested again. This process is repeated until a sufficient number of hits are obtained. Thus, iterative VS and HTS campaigns are designed to substantially increase the hit rates of random screening.

1.10.3 Filter functions. In addition to searching for active compounds, filter functions are very popular tools for VS. Filtering generally attempts to identify compounds with desired properties and discard others, and filter functions have been implemented for the analysis of rather diverse molecular properties including, for example, chemical reactivity, toxicity, drug-like character, or absorption, distribution, metabolism, excretion (ADME) parameters. It is also important to note that filter functions can have very different degrees of complexity. For example, NN or probabilistic QSAR models are often used as database filters but simple knowledge- or rule-based filters are also very useful. These filters are computationally easy to implement and very efficient.

The prediction of aqueous solubility of compounds is a good example for the diversity of filter functions. Here both QSPR and NN models have been used as database filters as well as molecular fragment-based or group contribution methods (Klopman and Zhao 2001) that calculate total solubility by adding up solubility values for listed molecular fragments. Passive absorption, the ability of molecules to pass through membranes, provides another example for a physico-chemical property that can be well predicted using a simple filter function. Two descriptors accounting for polar molecular surface area and log P(o/w), the octanol/water partition coefficient, are generally sufficient for accurate predictions and compounds with less than 140Å^2 polar surface area and a log P(o/w) between zero and four usually have favorable absorption characteristics (Egan *et al.* 2000). It is clear that such simple filter functions are efficient tools for screening of large databases and selection of compounds having acceptable or desired properties.

Both solubility and passive absorption belong to the spectrum of ADME parameters that play a major role in determining the *in vivo* efficacy of drug molecules. Other important effects include blood-brain-barrier penetration, metabolic stability (against cytochrome P450 isoforms, which are major drug metabolizing enzymes), serum protein binding, or oral availability. Essentially for all of these effects, computational models or rule-based filters have been established that can be used to screen databases (Selick *et al.* 2002). For predicting oral availability, the "rule-of-five" (Lipinski *et al.* 1997) has been formulated based on a statistical survey of drugs (Table 1.5). According to this rule, compounds have a good chance to be orally available, if they violate no more than one of four criteria.

In addition to property-related filter functions, structure-based filters play a major role in VS. Reactive or toxic group filters are based on dictionaries of undesired chemical moieties and are used to remove compounds from databases that are not suitable for medicinal chemistry programs (Walters *et al.* 1998). On the other hand, it is of course also possible to screen databases for compounds containing desired groups or pharmacophore arrangements (Muegge 2003). These types of filters are intuitive and widely applicable in synthetic and medicinal chemistry.

In pharmaceutical research, much emphasis has been put in recent years on the concept of "drug-likeness", which is based on the premise that specific structural features and/or molecular properties set drugs apart from other organic molecules. In fact, the rule-of-five is often cited in the context of drug-likeness and is, despite – or perhaps because of – its simplicity, credited with catalyzing many recent developments in this area. Both simulation methods and knowledge-based filters have been developed to mine databases for drug-like compounds (Clark and Picket 2000). Probably because it is to date not fully clear how drugs should "look like" and what their characteristic molecular features are, NN simulations have become popular to establish models to systematically distinguish between drugs and non-drugs that can be applied to filter databases. On the other hand, simpler knowledge-based techniques have also been developed that analyze structural features recurrent in drug molecules, compare (descriptor-based) property profiles of drugs and non-drugs, or monitor the presence of preferred pharmacophore arrangements in candidate molecules (Muegge 2003). Such approaches have been shown to detect a number of drugs with reasonable accuracy. Another recent trend has been to distinguish drug-likeness from lead-likeness (Rishton 2003). This idea takes into consideration that VS might rarely be capable of identifying mature drug candidates. Therefore, it should rather focus on the detection of "leads", i.e. molecules that can be chemically optimized to ultimately become drugs. Table 1.5 shows a comparison of simple drug-like and lead-like features that provide some guidelines for compound selection.

The comparison of these criteria was adapted from Rishton 2003. In the Drug-like column, the first four criteria represent to the rule-of-five.

However designed and implemented, the major purpose of compound filtering techniques is to enrich libraries with compounds having desired features, which can be accomplished by both positive and negative selection.

TABLE 1.5.   Drug-like versus lead like compound characteristics

Drug-like	Lead-like
MW < 500	MW < 350
ClogP < 5	ClogP < 3.0
Hydrogen bond donors < 5	Chemically stable
Hydrogen bond acceptors < 10	
Number of rotatable bonds $\leq$ 10	
PSA $\leq$ 140Å^2	
Peptides not suitable	Non-substrate peptides suitable
Eliminate reactive functional groups, promiscuous inhibitors, and metabolically unstable compounds	

1.11   FROM THEORY TO PRACTICE

A basic understanding of the concepts and methods reviewed thus far is critical for a meaningful application of many chemoinformatics software tools. The problem of how to best select and apply such methods for practical purposes should not be underestimated. The chemoinformatics literature is presently dominated by algorithm and method developments and reports of practical application or problem solving examples are still relatively rare. This can in part be explained by the situation that many applications in the pharmaceutical and chemical industry are proprietary. Thus, it is often considerably easier to publish methods than the results of "real life" projects. However, this also means that the current literature typically provides better answers to the question "what can be done?" than "what has been done – and how?", which does not make it easier for researchers interested in chemoinformatics to understand how to best tackle and solve their application problems. A major drawback of many commercially available computational tools is their "black box" character. In many instances, methodological and implementation details are hidden from the user, which – among other things – can make it very difficult to understand the limitations of the methodologies. However, the limitations of these tools is at least as important for successful applications as appreciating their opportunities, if not more so. What needs to be kept in mind is that for any given chemoinformatics problem, there is probably no ultimate methodological answer (just like it is not possible to design a generally applicable chemical descriptor space), and different approaches might often produce qualitatively similar results. It is often crucial to apply different techniques in parallel and pay particular attention to the reproduction of available experimental results and the formulation of experimentally testable hypotheses. For a practitioner in the chemoinformatics area, it is clearly very important to compare different computational tools and their functionalities and gain experience by studying test cases. However, being aware of potential caveats and shortcomings, practical chemoinformatics can not only make a significant impact on chemistry programs, but can also be "great fun". The following sections provide a few examples and case studies.

1.11.1   Database design.   The need to analyze, archive, and manage rapidly growing amounts of chemical is one of the major roots of chemoinformatics R&D. For these purposes, the design and implementation of different types of databases is of crucial importance. Fundamental to advanced database design is the concept of the "relational database" where objects and different types of information are linked and made accessible to the user in context. For example, in chemistry, a simple relational database could store compound structures, availability, and analytical data, and a more advanced version of this database might also include biological assay or activity data for these compounds. The use of relational chemical databases is one of the most important aspects involved in many chemoinformatics applications. Moreover, relational databases provide a convenient platform for interactions between computational and experimental scientists.

Briefly, in relational databases, diverse types of information are stored in different data tables that are connected and often embedded in an ORACLE database architecture

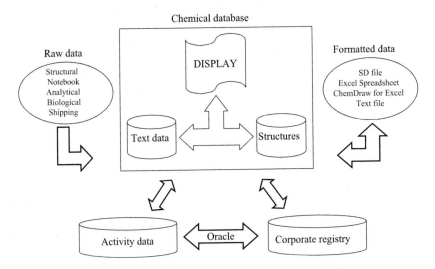

Figure 1.23. Schematic representation of relational database architecture (figures 1.23–1.25 were kindly provided by E. James Schermerhorn)

(http://www.oracle.com). Communication between different elements of the database is generally facilitated by using the Scientific Query Language (SQL). Typically, the user communicates with the database using "hidden" SQL through an interactive graphical interface to retrieve relational database information. An important feature of relational database design is that different types of data structures and software tools can be easily combined and transformed into coherent database architecture. This makes it possible to customize local databases for specific applications.

An instructive example of database customization has been provided by E. James Schermerhorn of Albany Molecular Research, Inc. (AMRI), who is engaged in the design and implementation of relational databases for various chemistry project teams to address their specific needs. Figure 1.23 schematically outlines the different types of data and database elements that typically need to be combined to support the work of synthetic or medicinal chemistry teams and document their efforts. Figure 1.24 shows a specific example of a relational database designed for a synthetic chemistry team. It displays the different data tables that represent the backbone of the database and the relationships that are established between them. This particular relational database connects synthesized compounds and their structures, analytical data, storage and shipment information, and notebook records. The central "MolTable" provides the link to compound structures through an identifier (Mol ID) and thus synchronizes textual and structural data. Established relationships are not necessarily equivalent but often hierarchical ("parent-child relationships"), which makes it possible to control and ensure data integrity. For example, a compound

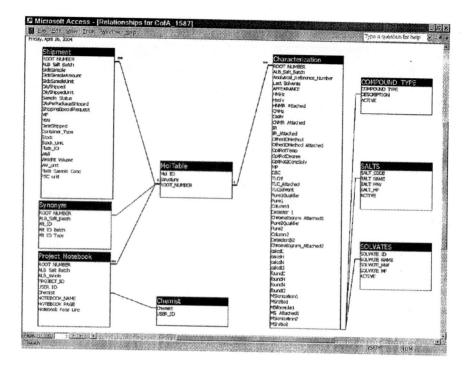

Figure 1.24.   Example of a relational database

structure needs to be drawn only once and can subsequently be linked to various types of data that are all stored in separate tables. Figure 1.25 displays the actual user interface of this database that was customized using the ChemFinder program (CambridgeSoft Corporation; http://www.cambridgesoft.com). Through this interface, a project chemist can input his molecules and chemical data, connect it to his notebook records, and retrieve other compound information. The next higher level of relational data base design could be, for example, the integration of contributions from different chemistry teams that collaborate on a larger project.

Clearly, much of the progress of collaborative R&D significantly depends on the availability of advanced and flexible database structures for data handling, sharing, and management. The design and implementation of such databases continue to be major tasks for chemoinformatics. This has also been realized by a number of software companies that provide commercial compound database and management systems and offer customization services.

1.11.2   Compound selection for medicinal chemistry.   Going beyond data analysis and management, chemoinformatics can form viable interfaces with medicinal chemistry and biological screening, as already mentioned before. In medicinal chemistry, chemoinformatics approaches support a diverse array of R&D activities,

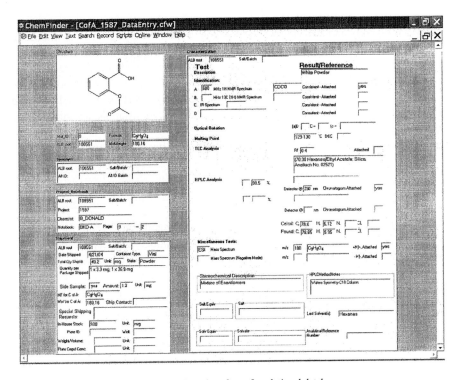

Figure 1.25. User interface of a relational database

ranging from the selection of compounds that are suitable for medicinal chemistry applications or the design of focused libraries to computational lead optimization efforts.

Michael S. Lajiness and colleagues, then at Pharmacia, have reported an interesting case study focusing on compound selection (Lajiness and Shanmugasundaram, 2004). These researchers provided a panel of 13 medicinal chemists with lists containing a total of 250 candidate compounds that were originally rejected by a senior investigator. However, the 13 chemists produced rather different sets of accepted and rejected compounds and consistently agreed in only very few cases. Moreover, dependent on how compound lists were arranged, each chemist rejected different sets of compounds from alternative lists. These findings clearly illustrated the limitations of subjective compound evaluation and selection, regardless of chemical knowledge and experience, and made a strong case for the application of more unbiased computational measures for compound selection. Accordingly, Lajiness and his colleagues developed a multi-component scoring system that made it possible to efficiently screen large source databases and select preferred compounds in a more unbiased and reproducible manner (Lajiness and Shanmugasundaram 2004).

CHEMOINFORMATICS

The following component scores were designed: a Q-score (QS) to assess general compound quality and lead-like potential (using various descriptors such as molecular weight, log P(o/w), solubility, or rule-of-five etc.), B-score (BS) summarizing known potency, selectivity, and toxicity of library compounds that had already been assayed in-house, D-score (DS) providing a measure of dissimilarity of a test compound from others in the library or selection set, and S-scores (SS) reporting the similarity of a compound to a selected template molecule (if available), for example, a known lead or drug candidate.

QS was implemented as a combination of different descriptor contributions:

$$QS = \frac{\sum_{i=1}^{n} w_i S(d_i)}{\sum_{i=1}^{n} w_i}$$

where w_i is the weight of $S(d_i)$, which represents the (regularized) score for the ith of n chosen descriptors. Dividing the sum of the resulting products by the sum of weights produces QS values between 0 and 1. Based on this formalism, descriptor values need to be transformed into component scores, as illustrated in Figure 1.26 for molecular weight (MW).

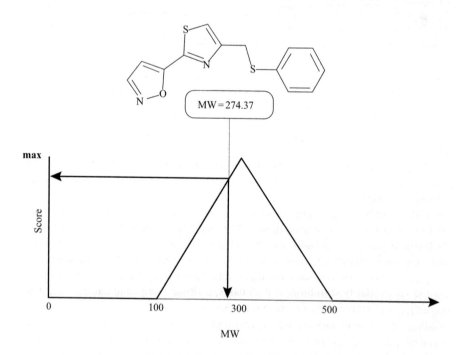

Figure 1.26. Example of a component score for a single descriptor (molecular weight, MW; adapted from Lajiness and Shanmugasundaram 2004)

Thus, using this scoring scheme a maximum weight score would be assigned for a compound with MW of 300 and compound with MW outside the preferred range (100–500) would obtain no score.

Furthermore, the biological score was defined as:

$$BS = \frac{(w_P S_{\text{Potency}} + w_S S_{\text{Selectivity}} + w_T S_{\text{Toxicity}})}{(w_P + w_S + w_T)}$$

After regularization of the component scores, the following consensus score (CS) was calculated:

$$CS = (w_{QS}QS + w_{BS}BS + w_{DS}DS + w_{ss}SS)$$

This type of consensus scoring scheme could be applied to quickly screen in-house libraries or other compound collections and rank molecules according to desired parameters. Although this multi-component score consists of a variety of individual scoring terms and associated weights, it is essentially a filter function. Importantly, this type of scoring scheme is variable and can be readily adjusted according to a chemist's needs. For example, an investigator may put high emphasis on compound quality and relative diversity during the early stages of a discovery project, whereas during later stages, the biological score or similarity to already discovered leads may be more emphasized. This type of computational filtering and ranking can process much more complex compound information in an organized manner than individual knowledge or chemical intuition would be able to. Thus, the work of Lajiness and colleagues provides a good example for the relevance of more objective than subjective compound selection schemes. Ultimately, this type of analysis depends on computations and the application of chemoinformatics-type approaches.

1.11.3 Computational hit identification. Virtual screening and its ability to complement HTS represent another attractive area for practical application of chemoinformatics methods. Instructive case studies were generated by us in collaboration with Florence L. Stahura. In a larger magnitude VS study, ligand-based virtual screening methods were applied to two confidential drug targets belonging to the G-protein coupled receptor superfamily. For these targets, a few active compounds had been identified in-house or reported by competitors and were used as search templates. As VS methods, two statistical partitioning algorithms were applied as well as a 2D-fingerprint for similarity searching. In both cases, a virtual compound database consisting of ~4.5 million compounds was screened.

Results obtained for target 1 are summarized in Figure 1.27. In this case, a total of 122 compounds were selected by VS calculations and assayed. Five new and structurally diverse hits were identified, all having low micromolar potency (which is a fairly typical level of potency observed for hits from VS).

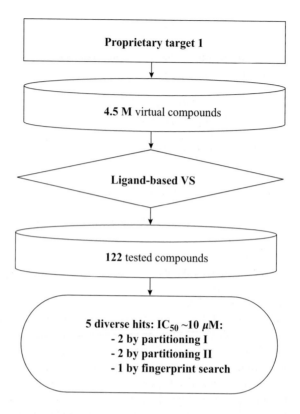

Figure 1.27.   Summary of a VS project. Partitioning I and II refer to two distinct partitioning methods
(figures 1. 27 and 1.28 were kindly provided by Florence L. Stahura)

Importantly, each of the different VS methods used here identified one or two distinct hits that were not detected by the other methods. These findings illustrate that parallel application of different types of VS methods typically increases the probability of success. This is due to the fact that the performance of VS methods is often influenced by compound class-specific features.

Target 2 presented an even more interesting and challenging test case. The results are summarized in Figure 1.28. Here a collaborator had already screened ~300,000 compounds prior to VS analysis, but could not identify any novel hits. Based on compounds patented by another pharmaceutical company, VS analysis was carried out following the unsuccessful HTS campaign. Out of 4.5 million virtual database compounds, only 30 candidates were selected and only 16 of these molecules could actually be acquired from chemical vendors. However, one of these 16 compounds that were assayed turned out to be a novel hit and was structurally distinct from the search templates. Obtaining such results from VS is certainly far from being routine. Selecting by computational means a very small number of test molecules from millions

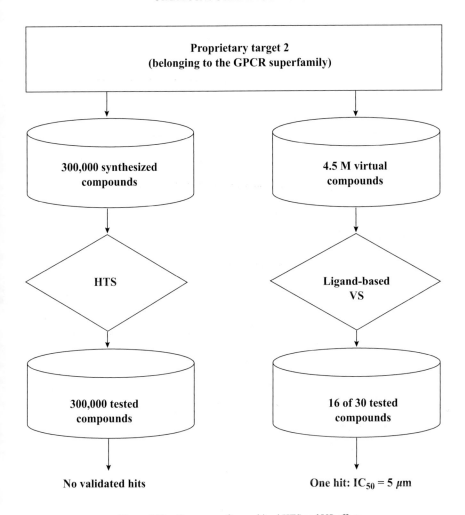

Figure 1.28.   Summary of a combined HTS and VS effort

of database compounds and, by doing so, identifying new actives is a task much like finding "needles in haystacks". It is rather fascinating that computational method-ologies can be successfully applied in such cases and, consequently, theses results should catalyze further progress in the chemoinformatics area.

REFERENCES

Agrafiotis DK (2002) Multiobjective optimization of combinatorial libraries. J Comput Aided Mol Des 16:335–356.

Agrafiotis DK, Rassokhin DN, Lobanov VS (2001) Multi-dimensional scaling and visualization of large molecular similarity tables. J Comput Chem 22:488–500.

Bajorath J (2002) Integration of virtual and high-throughput screening. Nat Rev Drug Discov 1:882–894.

Bajorath J (2004) Understanding chemoinformatics: a unifying approach. Drug Discov Today 9:13–14.

Brown FK (1998) Chemoinformatics: what is it and how does it impact drug discovery. Annu Rep Med Chem 33:375–384.

Clark DE, Picket SD (2000) Computational methods for the prediction of "drug-likeness". Drug Discov Today 5:49–58.

Cramer RD III, Redl G, Berkoff CE (1974) Substructural analysis: a novel approach to the problem of drug design. J Med Chem 17:533–535.

Cramer RD III, Patterson DE, Bunce JD (1988) Comparative molecular field analysis (CoMFA). 1. Effect of shape on binding of steriods to carrier proteins. J Am Chem Soc 110:5959–5967.

Egan WJ, Merz KM Jr, Baldwin JJ (2000) Prediction of drug absorption using multivariate statistics. J Med Chem 43:3867–3877.

Forrest S (1993) Genetic algorithms – Principles of natural selection applied to computation. Science 261:872–878.

Free SM Jr, Wilson JW (1964) A mathematical contribution to structure-activity studies. J Med Chem 7:395–399.

Gillet VJ (2002) Reactant- and product-based approaches to the design of combinatorial libraries. J Comput Aided Mol Des 16:371–380.

Gillet VJ, Khatib W, Willet P et al. (2002) Combinatorial library design using multiobjective genetic algorithm. J Chem Inf Comput Sci 42:375–385.

Gillet VJ, Willett P, Bradshaw J (2003) Similarity searching using reduced graphs. J Chem Inf Comput Sci 43:338–345.

Godden JW, Furr JR, Xue L et al. (2004) Molecular similarity analysis and virtual screening by mapping of consensus positions in binary-transformed chemical descriptor spaces with variable dimensionality. J Chem Inf Comput Sci 44:21–29.

Gund P (1977) Three-dimensional pharmacophore pattern searching. In: Hahn FE (ed), Progress in molecular and subcellular biology, vol 5. Springer-Verlag, Berlin, pp 117–142.

Hann M, Green R (1999) Chemoinformatics – a new name for an old problem? Curr Opin Chem Biol 3:379–383.

Hansch C, Fujita T (1964) r-s-p analysis. A method for the correlation of biological activity and chemical structure. J Am Chem Soc 86:1616–1626.

Hopfinger AJ (1980) A QSAR investigation of dihydrofolate reductase inhibition by Baker triazines based upon molecular shape Analysis. J Am Chem Soc 102:7196–7206.

Hopfinger AJ, Wang S, Tokarski JS et al. (1997) Construction of 3D-QSAR models using the 4D-QSAR analysis formalism. J Am Chem Soc 119:10509–10524.

Jarvis RA, Patrick EA (1973) Clustering using a similarity measure based on shared near neighbors. IEEE Trans Comput C22:1025–1034.

Johnson MA, Maggiora GM (eds) (1990) Concepts and applications of molecular similarity. Wiley, New York, 1990.

Kier LB (1997) Kappa shape indices for similarity analysis. Med Chem Res 7:394–406.

Kitchen DB, Stahura FL, Bajorath J (2004) Computational techniques for diversity analysis and compound classification. Mini Rev Med Chem 4:1029–1039.

Klopman G, Zhao H (2001) Estimation of aqueous solubility of organic molecules by the group contribution approach. J Chem Inf Comput Sci 41:439–445.

Kohonen T (1989) Self-organization and associative memory. Springer-Verlag, Berlin.

Labute P (1999) Binary QSAR: a new method for the determination of quantitative structure activity relationships. Pac Symp Biocomput 4:444–455.

Lajiness MS (1997) Dissimilarity-based compound selection techniques. Perspect Drug Discov Des 7/8:65–84.

Lajiness MS, Shanmugasundaram V (2004) Strategies for the identification and generation of informative compound sets. Methods Mol Biol 275:111–130, 2004.

Lewell XQ, Judd DB, Watson SP, Hann MM (1998) RECAP – retrosynthetic combinatorial analysis procedure: a powerful new synthetic technique for identifying privileged molecular fragments with useful application in combinatorial chemistry. J Chem Inf Comput Sci 38:511–522.

Lipinski CA, Lombardo F, Dominy BW, Feeney PJ (1997) Experimental and computational approaches to estimate solubility and permeability in drug discovery and development settings. Adv Drug Deliv Rev 23:3–25.

Martin EJ, Blaney JM, Siani MA *et al.* (1995) Measuring diversity: experimental design of combinatorial libraries for drug discovery. J Med Chem 38:1431–1436.

Martin YC (2001) Diverse viewpoints on computational aspects of molecular diversity. J Comb Chem 3:231–250.

Mason JS, Morize I, Menard PR *et al.* (1999) New 4-point pharmacophore method for molecular similarity and diversity applications: overview over the method and applications, including a novel approach to the design of combinatorial libraries containing privileged substructures. J Med Chem 42:3251–3264.

Muegge I (2003) Selection criteria for drug-like compounds. Med Res Rev 23:302–321.

Pearlman RS, Smith KM (1998) Novel software tools for chemical diversity. Perspect Drug Discov Des 9:339–353.

Rishton GM (2003) Non-lead-likeness and lead-likeness in biochemical screening. Drug Discov Today 8:86–96.

Roberts G, Myatt GJ, Johnson WP *et al.* (2000) LeadScope: software for exploring large sets of screening data. J Chem Inf Comput Sci 40:1302–1314.

Rusinko A III, Farmen MW, Lambert CG *et al.* (1999) Analysis of a large structure/biological activity data set using recursive partitioning. J Chem Inf Comput Sci 39:1017–1026.

Schnur D, Beno BR, Good A, Tebben A (2004) Approaches to target class combinatorial library design. Methods Mol Biol 275:355–377.

Selick HE, Beresford AP, Tarbit MH (2002) The emerging importance of predictive ADME simulation in drug discovery. Drug Discov Today 7:109–116.

Sheridan RP, Kearsley SK (2002) Why do we need so many chemical similarity search methods? Drug Discov Today 7:903–911.

Stahura FL, Bajorath J (2004) Virtual screening methods that complement HTS. Comb Chem High Throughput Screening 7:259–269.

Stanton DT, Jurs PC (1990) Development and use of charged partial surface area structural descriptors in computer-assisted quantitative structure-property relationship studies. Anal Chem 62:2323–2329.

Tetko IV, Kovalishyn VV, Livingstone DJ (2001) Volume learning algorithm artificial neural networks for 3D-QSAR studies. J Med Chem 44:2411–2420.

Todeschini R, Consonni V (2000) Handbook of molecular descriptors. In: Mannhold R, Kubinyi H, Timmerman H (eds), Methods and principles in medicinal chemistry 11. WILEY-VCH, Weinheim.

Walters WP, Stahl MT, Murcko MA (1998) Virtual screening – and overview. Drug Discov Today 3:160–178.

Ward JH (1963) Hierarchical grouping to optimize an objective function. J Am Stat Assoc 58:236–244.

Warmuth MK, Liao J, Rätsch G *et al.* (2003) Active learning with support vector machines in the drug discovery process. J Chem Inf Comput Sci 43:667–673.

Weininger D (1988) SMILES, a chemical language and information system. 1. Introduction to methodology and encoding rules. J Chem Inf Comput Sci 28:31–36.

Willett P (1987) A review of chemical structure retrieval systems. J Chemometrics 1:139–155.

Willett P (1988) Recent trends in hierarchic document clustering: a critical review. Inf Process Manag 24:577–597.

Xue L, Bajorath J (2000) Molecular descriptors for effective classification of biologically active compounds based on principal component analysis identified by a genetic algorithm. J Chem Inf Comput Sci 40:667–673.

Zupan J, Gasteiger J (1999) Neural networks in chemistry and drug design, 2nd edn. Wiley-VCH, Weinheim.

2.   PRACTICE AND PRODUCTS

2.1   ACCELRYS

I. Accelrys, http://www.accelrys.com
II. Accelrys is a computational science company that develops and delivers scientific software applications and services to solve R&D problems. They provide simulation and informatics software as well as a computational portfolio that combines modeling and visualization software to predict properties and interpret the behavior of molecules and materials and services. They adhere to an open, component-based software platform that can run across the network on Windows, Linux, or UNIX servers.
III. Key capabilities and offerings:
   a. **Modeling/Simulation Products** – Products such as *Materials Studio, Cerius², Insight II, Catalyst*, and *QUANTA* are the cornerstones of Accelrys' scientific computation products. Accelrys' newest platform, *Discovery Studio*. EDITOR'S NOTE: Given the *overwhelming number of product options* within each category, a brief topical description of the offerings will be provided. More detailed information is available on the website.
      i. *Materials Studio* – MS Modeling is a complete modeling and simulation product suite designed for structural and computational researchers to screen a wide variety of materials and process variables *in silico*. It provides simulation environments to create, view, and analyze molecules ranging from materials, simple chemicals to small molecules, crystalline materials, polymers, surfaces, and catalysts.
      ii. *Cerius²* – This modeling and simulation environment for SGI IRIX workstations and other servers provides tools for life science modeling and simulation applications. The environment has particular strengths in rational drug design and structure-based drug design.
      iii. *Insight II* – This multi-component package is a 3D graphical environment for molecular modeling. Its user interface enables the flow of data between a wide range of scientific applications. The package is specifically developed for applications in the life and materials sciences. This 49-module package includes topics ranging from ligand docking, binding site analysis, force field calculations to miscibility predictions and NMR programs. It is an extensive array of tools for the drug discovery scientist.
      iv. *Catalyst* – This component helps you to concentrate experimentation on likely active compounds by utilizing the relationship between chemical

51

structures and their biological activities. These tools allow you to investigate interactive pharmacophore relationships such as multiple mappings, multiple overlays, hypothesis scoring, and customized features. The program gives you access to complementary capabilities such as:

1. *generation of multiple conformations* with extensive coverage of conformational space
2. pharmacophore-based *alignment* of molecules
3. shape-based three-dimensional *database searching*
4. automated *generation of pharmacophore hypotheses* based on structure activity relationship (SAR) data.
5. The following modules allow the user to access a broad range of drug discovery information including *Catalyst/COMPARE, Catalyst/VISUALIZER, Catalyst/SHAPE*, and *ConFirm*.
6. Catalyst Analysis and Management
   (a) *Catalyst/INFO* builds and administers databases of 3D structures from your own data
   (b) *Catalyst Databases* provide hundreds of thousands of potential lead compounds, accessed through Catalyst's advanced information retrieval, analysis, and simulation capabilities. Databases include Derwent WDI, and BioByte Master File. In addition, DBServer comes pre-loaded with NCI databases and the Maybridge catalog.
   (c) *Catalyst DBServer* identifies leads by using a hypothesis/pharmacophore that may include binding features, structural and property constraints, as a search query against databases containing hundreds of thousands of molecules. DBServer considers 3D flexibility and comes preloaded with the Maybridge catalog and NCI databases.
7. Pharmacophore Identification using Catalyst
   (a) *HypoGen* generates 3D hypotheses to explain variations of activity with the structure of drug candidates.
   (b) *HipHop* performs feature-based alignment of a collection of compounds and generates pharmacophore models.

v. *Quanta* – This extensive library of crystallographic software programs streamlines and accelerates protein structure solution. The molecular graphics analysis program integrates a range of computational methods through a consistent interface. The following modules comprise this offering *X-POWERFIT, X-AUTOFIT, X-BUILD, X-LIGAND, CNX, X-SOLVATE, CHARMm, MCSS/HOOK* and *MMFF*.

vi. *Discovery Studio (DS)* – DS is a single unified platform of life science applications that enables discovery scientists to capture, manage, analyze, and use biological and chemical data and information within an integrated system. This desktop modeling, simulation, and informatics software platform combines modeling and visualization to analyze and predict the properties and behavior of chemical and biological systems.

The following is a summary of the product offerings within Discovery Studio: *DS Accord Biostore, DS Accord Chemistry Cartridge, DS Accord Enterprise Core, DS Accord for Excel, DS MedChem Explorer*, and *DS Gene* which is a sequence analysis package for Windows. Also for bioinformatics, proteomics, and ligand analyses the following products are available: *DS SeqStore, DS Modeling, DS GeneAtlas, DS AtlasStore, DS XBUILD, DS XLIGAND*, and *DS HT-XPIPE*.

b. **Informatics** – In the simplest terms, *informatics* is information technology focused on management and use of chemical and biological information, including databases and tools for data mining, data analysis, and decision support. Accelrys' focus on informatics includes:

   i. Bioinformatics – Bioinformatics is the set of technologies and tools used to store, retrieve, and analyze genomic and protein sequence data. These tools enable researchers to manage and manipulate sequence data, and to use that sequence data in the study of gene, protein, and DNA function.

      1. Accelrys provides data management and sequence analysis tools through the *DS SeqStore* and *GCG Wisconsin Package* products, running on UNIX systems. *DS Gene* provides PC access for visualization and analysis capabilities within the *Discovery Studio* product family.

   ii. Chemoinformatics – Chemoinformatics is the management and analysis of data that describes chemical compounds, for example, chemical structure and properties.

      1. Accord Enterprise Informatics (AEI) products integrate to offer an Oracle-based enterprise-wide solution for chemical information management. These products draw on the *power of the Accord Chemistry Engine* and Oracle databases to store, search, and analyze chemical structures, related biological and chemical data, experimental results, and registration information.

         (a) The Accord Enterprise Informatics system provides chemists with an informatics framework to store and analyze corporate chemistry data. The basic client-server chemistry system consists of a software collection (containing the default Oracle schema for chemistry, the data catalog and hit list manager) and uses the *DS Accord Chemistry Cartridge* to provide chemistry representation and searching services on the server side.

         (b) Accord Solutions for Chemistry is designed for chemists, screeners, pharmacologists and scientific software developers, who need to register, search, and share 2D molecular structures, reactions, and associated data.

         (c) Modules of the Accord Enterprise Informatics Suite include DS Accord Enterprise Core, the Data Catalog, List manager, the DS Accord Chemistry Cartridge, the DS Accord Registration, and Enumeration tools and the Accord Chemistry Engine that handles a wide range of organic, organometallic and inorganic molecules and reactions; isotopes, charges, radicals, delocalized pi-systems.

IV. Review:

Their software platform is open, component-based, and works with commonly used systems such as Windows, Linux, UNIX, and Oracle databases. The drug discovery product offerings cover an expansive array of topics from basic lead optimization to Lipinski factors and polymorph predictions. The Drug Discovery/ Bioinformatics platforms address problems product confronted by pharmaceutical and biotech researchers in various fields. Product offering includes:

- The Discovery Studio life science platform encompasses capabilities ranging from medicinal chemistry and functional proteomics data to electron density maps of protein-ligand complexes.
- Insight II is a 49-module molecular modeling package that examines ligand docking and force field calculations.
- Quanta's crystallographic software enables scientists other than crystallographers to access protein structure answers easily and, therefore early in the drug discovery process.
- DS Gene is a stand-alone sequence analysis package that operates on Windows.
- DS GeneAtlas analyzes protein sequence and identifies biochemical function.

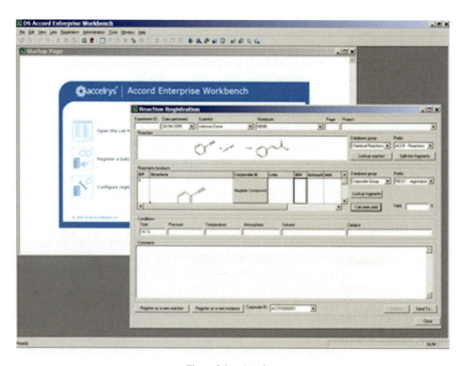

Figure 2.1.   Accelrys

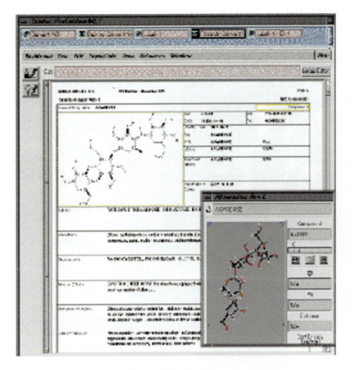

Figure 2.2.   Accelrys

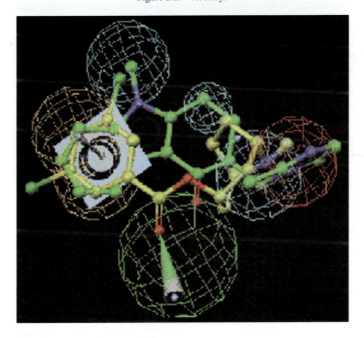

Figure 2.3.   Accelrys

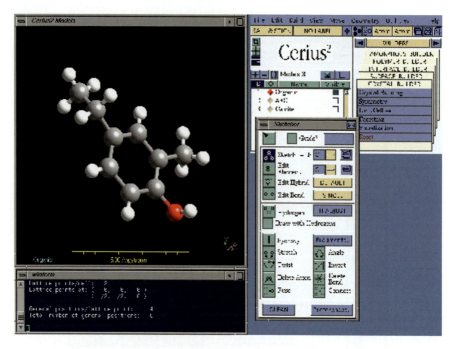

Figure 2.4.   Accelrys

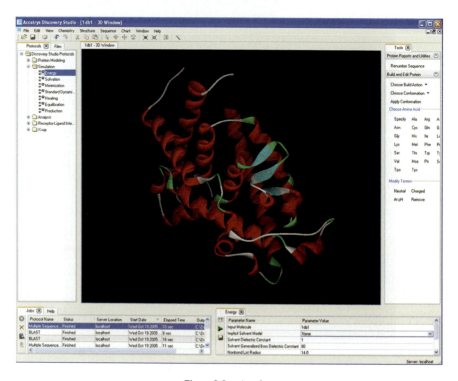

Figure 2.5.   Accelrys

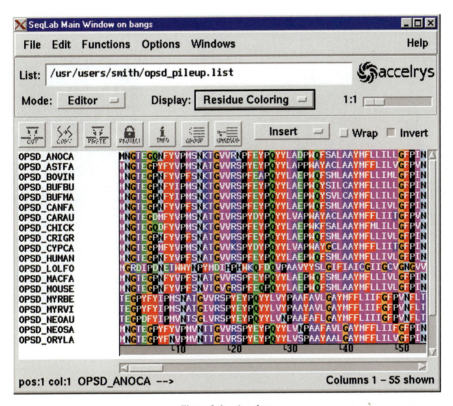

Figure 2.6.   Accelrys

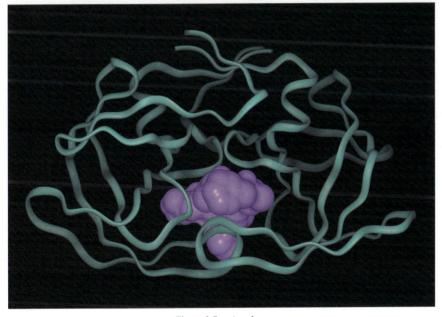

Figure 2.7.   Accelrys

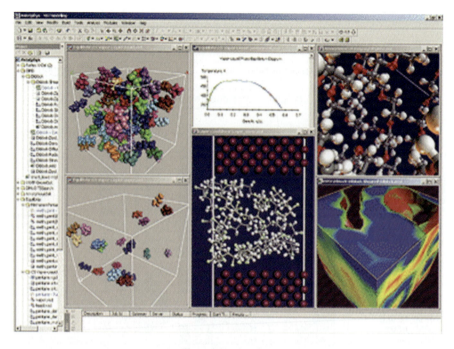

Figure 2.8.   Accelrys

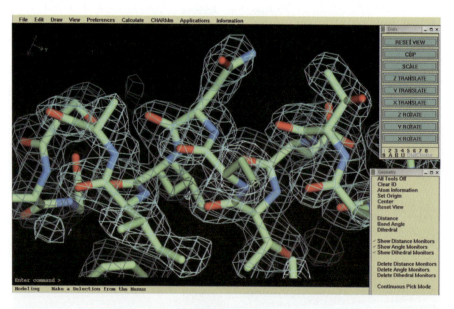

Figure 2.9.   Accelrys

2.2   ACD LABS

I. Advanced Chemistry Development Inc. (ACD Labs); http://www.acdlabs.com

II. Product Summaries:

a. **Software for Analytical Laboratory:** ACD/SpecManager product portfolio encompasses a number of technique-specific modules that handle a full range of analytical data (1D and 2D NMR, HPLC, LC/DAD, GC/IR, Mass Spectrometry, LC/MS, Infrared, Raman, UV-Vis-NIR, X-Ray Powder Diffraction, DSC, TGA, and a number of hyphenated data formats) from all major instrument vendors. All of the modules integrate to create a single platform for importing, processing, and reporting of analytical data, and enables creation of multi-technique, structure-searchable databases of experimental spectra. Numerous technique-specific spectral libraries are also available. In addition to desktop data processing and management, ACD/SpecManager Enterprise package utilizes the Oracle® database platform to enable global integration of analytical results and chemical structure information. ACD/Web Librarian offers visualization of the databases through a Web client browser interface to browse, review, utilize, and create reports. A number of supplementary software tools integrated to ACD/SpecManager aid in the analytical interpretation and identification of unknowns. Specifically, automation software, NMR and MS prediction, spectral libraries for verification and spectral matching, and expert tools for structure elucidation.

b. **Software for Chromatography Laboratory:** ACD/Labs offers ACD/Method Development Suite, which combines processing and databasing software (ACD/ChromManager) with integrated gas and liquid chromatography prediction modules (ACD/LC Simulator). A database of applications is included with the software, and is also available at http://www.chromdb.com. Extensive structure- and sub-structure searchable compilation of chiral separation methods is also available. ACD/ChromGenius allows determining the best method from a set of standard methods for a particular sample.

c. **NMR and MS Simulators/Predictors:** ACD/Labs offers several prediction modules for Nuclear Magnetic Resonance (NMR) applications: ACD/HNMR, ACD/CNMR, ACD/NNMR, ACD/PNMR, ACD/FNMR, and ACD/2D NMR Predictor, as well as extensive databases of assigned and reviewed experimental NMR values. ACD/MS Fragmenter provides possible fragmentation pathways for the compound of interest. Together with NMR processing software (ACD/1D NMR, ACD/2D NMR Processors, part of the ACD/SpecManager portfolio), all these modules facilitate structure-to-spectrum confirmation.

d. **Structure Elucidation Software:** ACD/Structure Elucidator helps determine the chemical structure directly from the experimental spectra of the unknown using a Computer-Assisted Structure Elucidation (CASE) algorithm and a large library of compounds and structural fragments with experimental chemical shift information. Structures are elucidated primarily using 1D and 2D NMR data and a molecular formula, in conjunction with a variety of complementary data that might be available (MS or infrared spectral data).

e. **Compound Molecular Property Predictors:** Available as a Suite and separately, a set of integrated modules calculate physicochemical properties, such as boiling point, Log P, polar surface area, Log D as a function of pH, solubility as a function of pH, pK_a, etc. ACD/Structure Design Suite builds on these predictions to suggest structural modifications that result in the desired physicochemical properties.

f. **ACD/ChemSketch Drawing Package:** This chemical drawing package, compatible with various chemistry file formats, is fully integrated and included with all ACD/Labs products and comes with several modules to expand its capabilities (e.g., ACD/Dictionary, ACD/Tautomers, ACD/3D Viewer, and more).

g. **ACD/ChemFolder Databasing Package:** Desktop database management software for chemical structures, chemical properties, experimental data, and reaction mapping. It can work with Personal Digital Assistant (PDA) handhelds, and includes ACD/ChemCoder module to produce and scan 2D barcodes.

h. **Nomenclature Software:** Comprehensive software that offers nomenclature generation and name-to-structure conversion in compliance with IUPAC, CAS, and IUBMB rules, as well as the InChI protocol. Supports English, German and French languages, and a variety of naming variations.

III. Key Capabilities and Offerings:

    a. **Software for Analytical Laboratory:**

        i. ACD/SpecManager software encompasses a number of modules for analytical data import and processing, a report editor (ACD/ChemSketch), and a databasing component for storage and retrieval of disparate forms of analytical information. All modules, each of which can be used separately, integrate into a single master interface that automatically provides appropriate technique-specific expert tools, such as spectral interpretation, baseline correction, and others, according to the context of the selected data set. The databasing module lets users browse and mine the database content using a variety of spectral and structure-based search criteria. It also allows viewing of several analytical results and associated chemical structure information in a single interface, and reporting of data retrieved through structural, spectral, or textual queries. Technique-specific processing modules are:

            i. ACD/1D NMR Processor.
            ii. ACD/2D NMR Processor.
            iii. ACD/ChromProcessor.
            iv. ACD/Curve Processor.
            v. ACD/MS Processor.
            vi. ACD/UV-IR Processor.

       ii. A number of reference spectra databases and structure-spectrum correlation libraries are also offered. More details are available at the ACD Labs web site: http://www.acdlabs.com/products/spec_lab/exp_spectra/spec_libraries/

    iii. ACD/SpecManager Enterprise enables multiple users to work with the database simultaneously through client-server architecture. Available as:
- i. ACD/MS Manager Enterprise
- ii. ACD/1D NMR Manager Enterprise
- iii. ACD/2D NMR Manager Enterprise
- iv. ACD/UVIR Manager Enterprise
- v. ACD/Curve Manager Enterprise
- vi. ACD/ChromManager Enterprise
- vii. And a combination of the above modules

    iv. ACD/Web Librarian offers a Web-browser based interface to all ACD/Labs' databases of structures and their associated spectral, chromatographic, and curve data.

    v. ACD/Automation Server is designed to operate fully unattended, recognizing instrument and data types and executing a series of predefined processing tasks to offer automatic handling of many analytical processing, storage, and reporting tasks that can be qualified as routine, batch, or requiring minimal user intervention.

    vi. ACD/1D NMR Expert and ACD/2D NMR Expert provide a software platform for the high-throughput processing and databasing of 1D and 2D NMR data.

b. **Software for Chromatography Laboratory:**

    i. ACD/Method Development Suite uses an archive of successful separations and HPLC/GC modeling to design and develop separation methods. The Suite includes:
- i. ACD/ChromManager, which allows chromatographers to analyze and process experimental chromatograms, and build databases of successful separations, structures, and other method information. Includes the ACD/Chromatography Applications Database with thousands of publicly available separation methods.
- ii. ACD/LC Simulator, which simulates HPLC and GC separations and helps optimize the experimental conditions according to numerous parameters. This software also predicts pK_a values of the analytes from a structure, and allows pH optimization.

    ii. ChirBase Chiral Applications Databases for ACD/ChromManager are available for HPLC, GC and CE. HPLC-only database contains method information for over 100,000 HPLC experiments, including data for 30,000 unique structures linked to over 1,400 unique chiral selectors.

    iii. ACD/ChromGenius allows users to select the best method from a standard set of separation methods, based on the chemical structure of the analyte(s). Both optimal resolution and retention time parameters are also defined by the user.

c. **NMR and MS Simulators/Predictors:**

    i. ACD/HNMR: Assists chemists in the interpretation of ^{1}H NMR spectra by generating a predicted spectrum and providing predicted values for both chemical shifts and coupling constants for almost any organic chemical structure. An extensive database of assigned ^{1}H NMR shifts is available.

ii. ACD/CNMR: Aids in the interpretation of ^{13}C NMR spectra by predicting ^{13}C chemical shifts and coupling constants. The software also generates a predicted ^{13}C NMR spectrum that can be compared directly to the experimental spectrum. An extensive database of assigned ^{13}C NMR shifts is available.

iii. ACD/NNMR: Predicts chemical shifts and coupling constants to assist in interpretation of ^{15}N spectra and for setting up the experimental conditions when acquiring ^{1}H-^{15}N heteronuclear shift correlation data. A database of assigned ^{15}N NMR shifts is available.

iv. ACD/PNMR: Predicts chemical shifts and coupling constants to aid in the interpretation of ^{31}P NMR data. A database of assigned ^{31}P NMR shifts is available.

v. ACD/FNMR: Calculates ^{19}F chemical shifts and coupling constants for chemical structures containing fluorine. A database of assigned ^{19}F NMR shifts is available.

vi. ACD/2D NMR Predictor: Calculates 2D spectra for a variety of experiments of several correlation types in heteronuclear or homonuclear environments. These include COSY, TOCSY, HSQC, HMBC, and more.

vii. ACD/MS Fragmenter: Generates molecular fragments according to a chosen mass spectrometry ionization type and fragmentation rules.

d. **Structure Elucidation Software:** ACD/Structure Elucidator combines the use of legacy data and ACD/Labs' Computer-Assisted Structure Elucidation (CASE) algorithm to predict and elucidate chemical structures of unknown chemical entities. The package also includes:

i. ACD/HNMR Predictor

ii. ACD/CNMR Predictor

iii. ACD/1D NMR Manager

iv. ACD/2D NMR Manager

v. ACD/UV-IR Manager

vi. ACD/MS Manager

e. **Compound Molecular Property Predictors:**

i. ACD/pK$_a$ DB: Available in batch and desktop versions, this module calculates acid-base ionization constants (pK$_a$ values at 25°C) and zero ionic strength in aqueous solutions for almost any organic molecule. Predicted values are provided with a detailed report on how the data was derived including Hammett-type equation(s), substituent constants, and literature references where available. Software package also includes an extensive structure-searchable database of experimental pK$_a$ values.

ii. ACD/Log P DB: Calculates the octanol-water partition coefficient (log P) for a wide range of neutral chemical compounds. Each calculation is provided with its 95% confidence interval. Also calculates polar surface area, "rule-of-5" parameters, and other related properties. Software package includes a structure-searchable database of experimental log P values. Available in batch and desktop versions.

   iii. ACD/Sigma: Allows chemists to directly access the electronic substituent constant, σ, calculated for selected fragments of their molecule or the substituents of a library. Available in batch and desktop versions.

   iv. ACD/Log D Suite: Calculates the distribution coefficient, log D (the pH-dependent log P), and related properties such as the bioconcentration factor (BCF) and organic carbon adsorption coefficient (K_{oc}) at any pH. The Suite includes all of the above predictive modules. Available in batch and desktop versions.

   v. ACD/Solubility DB: Calculates the aqueous solubility at any pH, as well as intrinsic solubility and solubility of the chemical dissolved in pure (unbuffered) water. Software package includes a structure-searchable database of experimental solubility values. Available in batch and desktop versions.

   vi. ACD/Structure Design Suite: This software package helps medicinal chemists choose substituent modifications and evaluate the physicochemical properties of the new analogs in order to optimize their bioavailability. Includes a substituent database containing a compilation of over 10,000 popular organic chemical substituents (neutral, acidic, basic substituents, and heterocycles).

   vii. ACD/Boiling Point: Calculates the boiling point, the enthalpy of vaporization at the boiling point, and the flash point of a compound.

f. **ACD/ChemSketch Drawing Package**, available as a separate product and included with the majority of ACD/Labs software products, offers numerous capabilities:

   i. Draw chemical structures, reactions and schematic diagrams, and produce professional reports and presentations that can be transferred to any OLE-supported software (e.g., Microsoft Office and Adobe Reader applications).

   ii. Support a broad variety of bond types and chemical classes. Import chemical structures. Calculate reactant quantities.

   iii. Generate InChI identifiers from structures, and vice versa.

   iv. Search by structure in Microsoft Office, Adobe, or ACD/Labs documents.

   v. Generate 3D structure representations.

   vi. Determine tautomeric forms of a compound.

   vii. Find the chemical structure of a compound by its trivial name.

   viii. Generate IUPAC names of chemical structures (restricted version).

   ix. Integrate with third-party software (MDL ISIS/Draw and ISIS/Base/Host; CambridgeSoft's ChemDraw among others).

   x. Transfer structures to a PDA device.

   xi. Generate 2D barcodes for structures.

xii. Use ACD/ChemSketch as an interface to the ACD/Labs Online Service.

g. **ACD/ChemFolder Databasing Package:**

    i. Structure-enabled data management software capable of handling chemical structures (including "generic" or Markush structures), multi-step reactions, biotransformation maps, chemical properties, images, text, experimental data, hyperlinks, and more. Can hold up to 512,000 records with up to searchable 16,000 data fields each, and it can provide advanced graphic presentation and statistical analysis of stored data.

    ii. Includes software for data transfer to the PDA handhelds.

    iii. Includes ACD/ChemCoder module to produce and scan 2D barcodes.

    iv. Includes ACD/ChemSketch software.

h. **Nomenclature Software:**

    i. Basic IUPAC naming and trivial name look-up are offered through ACD/ChemSketch.

    ii. ACD/Name Chemist Version allows chemists to obtain IUPAC names for almost any organic structure and selected classes of biochemical, organometallic, and inorganic structures. This program also produces chemical structures from systematic, trade and trivial names, and registry numbers.

    iii. ACD/Name offers nomenclature generation and name-to-structure conversion in compliance with IUPAC, CAS, and IUBMB rules, as well as the InChI protocol. Supports English, German and French languages, and a variety of naming variations. Available as a batch version.

IV. Review:

The integration of ACD/Labs' modules with third party software, drawing packages, and instrumentation makes this platform useful for a wider range of scientists. ACD/ChemSketch provides a drawing package that allows scientists to manage structural data applied in several research fields. Its tautomer and dictionary modules are handy for the medicinal chemist, and the integration of ACD/Labs' modules with CambridgeSoft's ChemDraw and MDL ISIS/Draw and ISIS/Base/Host gives the end user the option to use/pick the platform with which she/he is more comfortable working. The porting of the modules to be used with Personal Digital Assistants comes as a handy alternative for the mobile use. For educators and academia, ACD/Labs also has available a free-of-charge version of ChemSketch (freeware) containing several of the capabilities of the full version including systematic IUPAC name generation for small molecules and conversion of SMILES strings to structures and vice versa. The Chromatography Laboratory modules aid in the development of GC and LC optimal parameters for the separation of organic compounds in a mixture where the introduction of experimentally determined values increases the accuracy of these modules. Analytical data management modules that, in addition to the basic processing and storage, also include advanced chemometric tools can be paired with spectral predictions to enable spectral confirmation and elucidation.

A comprehensive set of Compound Molecular Property Predictors aid in Structure-Property Relationship elucidation and drug design. The capability of running calculations in batch mode allows the scientist to cover and explore both real and virtual sets of compounds. ACD/Labs' Tautomers module, despite its limitations (metal-containing structures, charged atoms, structures with coordinating bonds, 255-atom limit), addresses an important aspect of compound representation often times overlooked. It is also noted that the ACD/Labs product line does not include a docking module, a very useful tool that would aid the scientist in studying and proposing more enriched lead candidates. Also, none of the desktop packages is available for other OS such as the Mac OS X and Linux. However, the ACD/Physchem Batch and Name Batch packages are indeed available also for Linux.

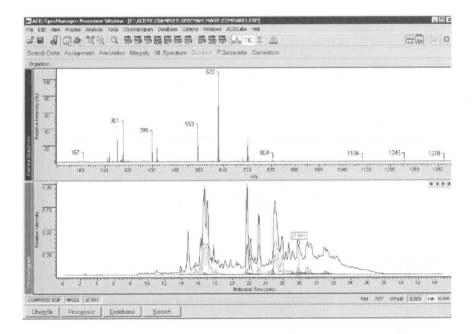

Figure 2.10.   ACD

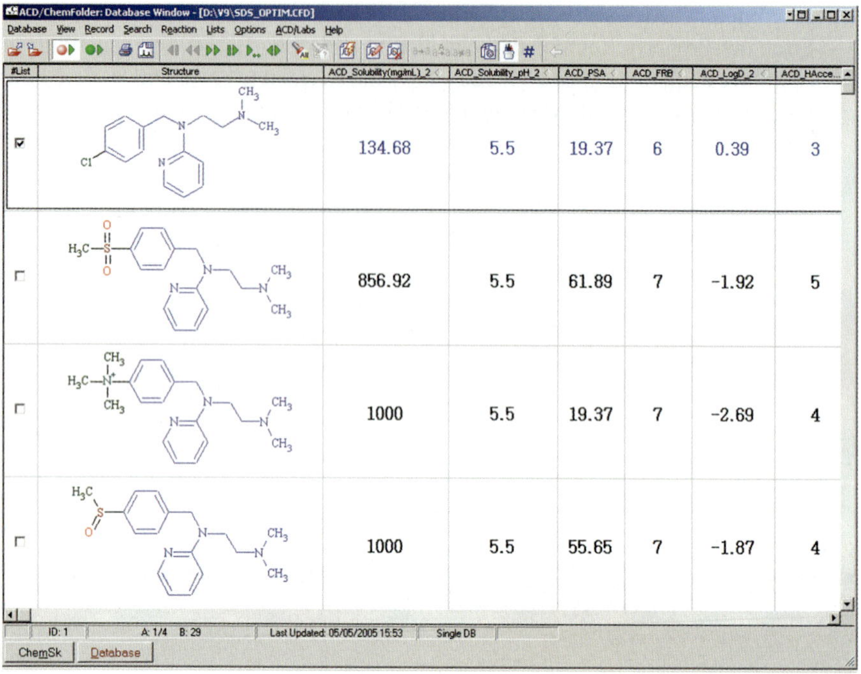

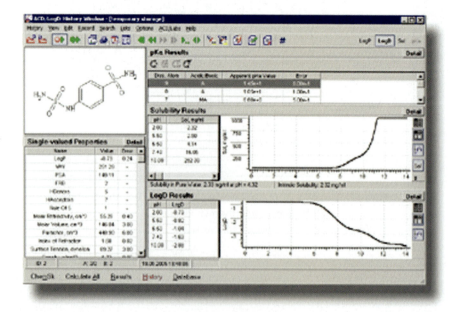

Figure 2.11.   ACD

Figure 2.12.   ACD

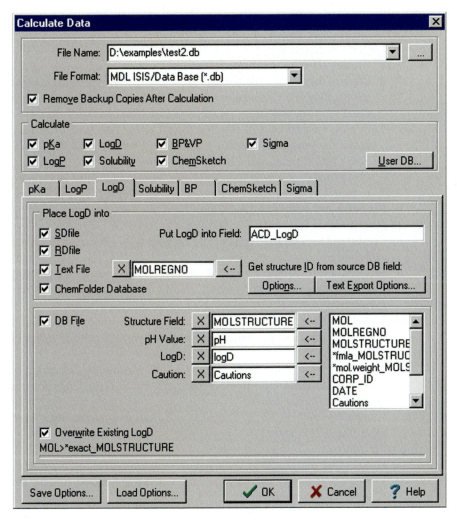

Figure 2.13. ACD

2.3 BARNARD CHEMICAL INFORMATION LTD

I. Barnard Chemical Information Ltd. (BCI); http://www.bci.gb.com/. As of 23rd September 2005, BCI became a wholly-owned subsidiary of Digital Chemistry Ltd. (http://www.digitalchemistry.co.uk), to whom all correspondence should be addressed.

II. **Product Summaries:** BCI specializes in software for analysis of the structural diversity of large chemical dataset and combinatorial chemistry libraries. BCI's software covers: Chemical Structure Fragments and Fingerprinting, Diversity Analysis, Cluster Analysis, and Structure or Reaction query format conversion (MOLSMART).

III. **Key capabilities and offerings:**
   a. *Chemical Structure Fragments and Fingerprinting.*
      i. Fingerprint generation.
         1. Creates 2D-structural fingerprints.
         2. Imports data in SD file format, MDL molfile format, SMILES text strings, Daylight Chemical Information Systems Inc.'s Thor Data Tree (TDT) format, and Tripos Inc.'s Sybyl Line Notation (SLN) format.
         3. Exports fingerprint data as a BCI-format screen file, Accelrys Binary data file, Daylight Chemical Information Systems Inc.'s Thor Data Tree (TDT) format, and a binary matrix file.
         4. Covers and uses several fragment types such as augmented atom, atom sequence, atom pair, and ring fusion among others.
      ii. Fragment and dictionary generation.
   b. *Diversity Analysis.*
      i. Creates and uses centroid and modal fingerprints for data set analysis and subset selection.
      ii. Includes the program CAMFER (Centroid And Modal Fingerprint ExtractoR) to generate and output centroid and generalized modal fingerprints.
      iii. Includes the program DIVINER (DIVersity INformation ExtractoR) to analyze centroid and generalized modal fingerprints to provide diversity information.
      iv. Includes the program RAMADER (RApid MAximum Dissimilarity Extractor) to extract a subset of most dissimilar compounds from a selected compound dataset.
      v. Compatible with Daylight Chemical Information Systems Inc.'s Thor Data Tree (TDT) format.
   c. *Cluster Analysis.*
      i. Handles and analyzes large chemical datasets.
      ii. Provides the Hierarchical Agglomerative Clustering package utilizing the Ward's and group-average clustering methods.
      iii. Provides the Divisive K-Means Hierarchical Divisive Clustering package.
      iv. Provides the K-Means Non-hierarchical Clustering package.
      v. Provides the Jarvis-Patrick Non-hierarchical Clustering package.
   d. *Structure or Reaction query format conversion (MOLSMART).*
      i. Converts MDL Molfiles (structure files) or MDL RGfiles and Rxnfiles (reaction files) to Daylight Chemical Information Systems Inc. SMARTS or SMIRKS strings.
   e. *BCI Toolkit.*
      i. The functionality of most of the methods listed above is available in toolkit form for calling from within a user's own programming environment.
      ii. In addition to the above, a *Markush toolkit* for handling combinatorial libraries is also available. This utilizes Markush analysis techniques

to enable the generation of SMILES, calculated physicochemical properties, topological indices and structural fingerprints for each item within a library, and enables full, substructure and overlap searching of libraries, all without enumeration of the individual structures. The Markush methods are several orders of magnitude faster than the equivalent methods applied to enumerated structures.

IV. Review:

BCI's software is aimed directly to chemical information systems for the processing, study and analysis of large data sets derived from historical compound databases, corporate databases, combinatorial chemistry, high-throughput synthesis, and high-throughput screens. The compound clustering and fingerprint techniques are of great value to medicinal chemists and combinatorial chemists for QSAR and virtual screening of both real and virtual compound libraries to pick candidates for further development. MOLSMART offers a powerful tool to execute faster and more efficient structure and reaction searches. BCI's software is designed to be handled primarily by computational chemists and software developers since the software is to be used in conjunction with chemical information systems. Porting some of these packages as stand alone PC products would allow end users to analyze and work on early and more advanced focused projects alleviating corporate IT's and computer power's bottlenecks. Note that, due to the utility nature of the BCI software (they are add-ons to a user's chemoinformatics GUI, and once launched they work in the background), screen shots are not available.

2.4 BIOBYTE

I. BioByte, http://www.biobyte.com

II. Product Summaries: BioByte provides a variety of software solutions to the drug discovery scientist that range from a chemically-oriented database system to a unified driver for ClogP/CMRprograms.

III. Key capabilities and offerings:

a. C-QSAR Package is their premier offering and is a comprehensive stand-alone drug discovery system. A brief summary of capabilities and program components is given below.

   i. QSAR – A regression program for drug designers (without extensive experience in statistics) that handles linear, bi-linear and parabolic equations with various variable transformations. The data entry method and procedures for verification of structures and parameters are user-friendly, and produce a variety of 2-D graphs as output.

   ii. BIO/PHYS – Dual databases of QSAR equations relating bio- and physicochemical activities to structural parameters. BIO currently contains – more than 12,700 equations for different biological systems, which includes anti-cancer agents, anti-HIV agents, anti-bacterials, topoisomerase inhibitors, COX inhibitors, and QSAR for ADME (mainly absorption, metabolism, etc.).

Users will find these valuable for validation of new equations, as they are being developed. Specifically, researchers examine emerging structure-activity relationships for resemblance to others with known mechanisms.

iii. SIGMA – A THOR database of 4,300 substituent structures with up to 40 electronic and/or steric parameter types and 20,000 values, fully referenced. Preferred parameters are listed, and automatic loading to QSAR is a time-saving feature of C-QSAR.

iv. BioLoom is the latest extension of Clog P that was offered earlier to calculate log P, CMR etc. The following features make it a very unique, simple and useful preliminary literature survey tool:

1. BioLoom, has more than 58,000 molecular structures.
2. It also accesses synonyms which makes the number ~ 77,000
3. Over 60,000 measured Log P/log D values in different solvent systems, (mostly in octanol water).
4. About 14,000 measured pKa values.
5. More than 32,200 CAS numbers.
6. More than 30,000 activity entries are listed under more than
7. 1,500 different activity types; e.g. Target as HIV protease inhibitor, and result as Anti-HIV.
8. Also now it is linked to QSAR database of Prof Corwin, which has more than 12,700 QSAR models using Hammett type parameters with their biological activity.

v. QSAR models are derived for different biological systems, which includes anticancer agents, anti-HIV agents, anti-bacterial, topoisomerase inhibitors, COX inhibitors, QSAR for ADME, mainly absorption, metabolism, etc.

vi. Each QSAR includes data such as reported biological activity, the physicochemical parameters, and structures stored in SMILES notation (canonicalized) etc.

vii. Currently the QSAR database has about 90,000 unique structures, and more are being added daily from sources such as Journal of Medicinal Chemistry, European Journal of Medicinal Chemistry, Journal of Biochemistry, Chemical Pharmaceutical Bulletin, Journal of American Chemical Society, Nature, Journal of Molecular Pharmacology, Quantitative Structure-Activity Relationship, and Journal of Pharmaceutical Sciences.

viii. More than 16,000 molecules have been linked from this database into the BioLoom to make its access easy to make various types of searches such as activity value, activity type, parameter value, references etc.

This is a database of more than 200,000 molecules, with their measured biological activity, and QSAR/equations derived internally, (at present there are more than 12,500 QSAR), using QSAR program developed by Dr. Corwin Hansch and his group.

b. Super Package is their next product offering. A brief summary is given below.

   i. THOR is a chemically oriented database management system with open-ended data-type entry that has read/write capability, and is capable of efficiently handling large proprietary databases (i.e. >500,000 compounds.).

   ii. GENIE is a powerful substructure specification language for efficient pattern recognition; the target language, SMARTS, is a simple extension of SMILES.

c. Regular Package is a system that efficiently handles large batch Clog P calculations.

   i. MASTERFILE is a THOR database that contains over 56,000 structures, 75,000 names, over 60,000 log P values in various solvent systems (including about 13,000 high-confidence octanol/water values, called LOGPSTAR), 14,000 pKa values, 32,200 CAS numbers, and much more.

   ii. MERLIN is a program which searches large databases for the presence of target substructures; returns 'hits' on files of over 100,000 structures in average of one or two minutes.

   iii. BDRIVE is the component that allows you to calculate ClogP values in very large batches.

d. Economy Package components are briefly described below.

   i. UDRIVE is a unified driver for the ClogP/CMR programs. From structural input via SMILES, this delivers a calculated logP octanol/water and molar refractivity value with error estimate and calculation details.

   ii. SMILES: Simplified Molecular Input Line Entry Specification, a simple, easy to learn, yet comprehensive chemical nomenclature, used in all of BioByte's programs.

IV. Review:

BioByte product offerings are diverse, easy to learn, yet powerful software packages for researchers, academics, and students. BioLoom, their next generation ClogP package, calculates hydrophobic and molecular refractivity parameters, but now is linked to their database which includes over 60,000 measured log P values, as well as 14,000 pKa's. A demo version is available for download.

Interesting information on C-QSAR is provided on their website through these links.

   Chem-Bio Informatics and Comparative QSAR

   C-QSAR: A General Approach to the Organization of Quantitative Structure-Activity Relationships in Chemistry and Biology

The biological activity searches are now much more powerful than before and are searchable by *activity types*. This allows researchers to narrow searches to provide just the relevant literature references.

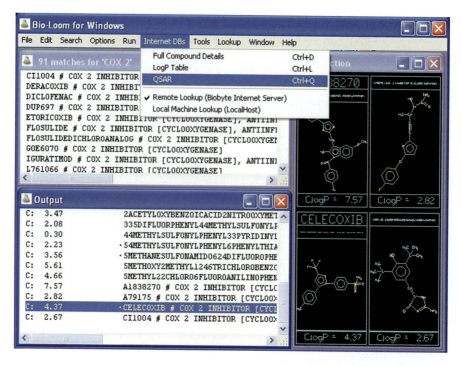

Figure 2.14.   Biobyte

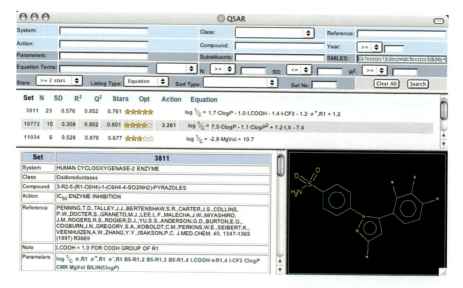

Figure 2.15.   Biobyte

2.5   CAMBRIDGESOFT

  I. CambridgeSoft, http://www.cambridgesoft.com
 II. Product Summaries:
     a. **Drawing packages: ChemDraw(9.0):** http://www.chemdraw.com ChemDraw
        is a widely used drawing package that is used to draw molecules, reaction
        schemes, and textual information in publication quality form. Information can
        easily be cut and pasted into word documents and other applications. Molecules
        can be converted into other molecular file formats and analyses (such as the
        theoretical calculation of ^1H and ^{13}C NMR spectra) are routine.
     b. **Chemistry working environment, Chemistry databases: ChemOffice**
        http://www.chemoffice.com ChemOffice consists of ChemDraw (drawing
        package), Chem3D (modeling package), E-Notebook (archive and databases),
        ChemFinder, ChemInfo and BioAssay for chemical publishing, modeling, and
        database work. ChemINDEX 9.0 contains NCI & AIDS data in DVD format.
        Oracle or SQL support allows organizations to share data, located by text or
        substructure searching, as well as maintain rigorous security and efficient
        archiving. ChemFinder is a chemically intelligent database manager and search
        engine that creates Excel-searchable spreadsheets. ChemFinder connects to
        Oracle and MS Access data and imports/exports MDL RD and SD files.
     c. **Cambridgesoft Enterprise Solutions:** Chemistry working environment &
        registration system http://www.cambridgesoft.com/solutions/ Enterprise
        Solutions allows the creation of internal databases and capturing of informa-
        tion in electronic notebooks programs.
     d. **Cambridgesoft Databases:** http://www.cambridgesoft.com/databases/Chem
        Finder.Com provides a full range of research databases and provides web
        links for over 100,000 chemicals that one can search by name, partial name
        or view structure drawings. Different computational engines can be config-
        ured to deliver the computed properties and large jobs can be diverted to net-
        work computational resources.
III. Key capabilities and offerings:
     a. **Drawing packages: ChemDraw (9.0):** *ChemDraw Ultra 9.0:*
          i. ChemNMR Enhanced provides Proton NMR prediction (with proton-
              proton splitting patterns and Carbon-13 shift values) and generates sys-
              tematic names for chemical structures.
         ii. Mass Fragmentation Tool examines potential mass fragments by break-
              ing bonds with the Mass Fragmentation tool.
        iii. MS Word Numbering assigns reference numbers to ChemDraw struc-
              tures that appear in your MS Word documents to use as a reference.
         iv. Polymer Draw allows one to represent and manipulate polymers in
              ChemDraw.
          v. Structure CleanUp improves poor drawings.
         vi. Online Menu allows one to draw a structure or model and immediately
              get online vendor information from ChemACX.Com.

    vii. ClogP property server calculates n-octanol/water partition coefficients.

   viii. ChemProp/Draw allows one to compute physical properties such as Log P, BP, and MP.

    ix. ChemINDEX Ultra 9.0 DVD Edition provides 4 Databases on 1 DVD:

      1. NCI Database – Over 200,000 compounds with anti-cancer drug dose-response data.

      2. AIDS Database – NCI compiled database for AIDS anti-viral compounds. *ChemINDEX 9.0* – Small molecule physical property data on over 70,000 compounds.

      3. ChemRXN 9.0 – Organic reaction databases include ChemSelect from Infochem GmbH and a selection from ISI's ChemPrep, for a total of over 29,000 reactions.

      4. *ChemDraw ActiveX/Plugin Pro 9.0* allows one to query online databases and view and publish online structures.

b. **Chemistry working environment, Chemistry databases: ChemOffice** http://www.chemoffice.com

      i. ChemDraw Spotfire allows one to view structures within DecisionSite from ChemFinder or Registration System to integrate chemical and biological data, and perform substructure or text searching. It requires Spotfire DecisionSite.

     ii. ChemSAR is a Chem3D Windows add-in for MS Excel with descriptive statistics and plots for structure-activity relationships.

    iii. Purchase/Excel creates a shopping list of compounds in Excel for automated chemical purchasing.

    iv. ChemDraw/Excel displays and performs calculations on up to 1,400 chemical structures in Excel.

     v. CombiChem/Excel uses ChemFinder for MS Excel to build combinatorial libraries with embedded ChemDraw structures. Other commercially available programs from Afferent/MDL, Accord SDK/Accelrys, LUCIA/Sertanty, and Pipeline/Scitegic can also enumerate actual and virtual libraries.

    vi. Included databases are as follows:

      1. ChemINDEX CD-ROM (180,000 compounds from NCI)

      2. ChemMSDX CD-ROM (7,000 material safety datasheets)

      3. ChemRXN CD-ROM (29,000 organic reactions)

      4. *The Merck Index 13th CD-ROM Edition* is an encyclopedia of chemicals, drugs and biologicals with over 10,000 monographs on single substances or groups of related compounds.

      5. ChemACX CD-ROM (300 catalogs for ordering over 450,000 products)

      6. ChemACX-SC CD-ROM (leading screening compound suppliers)

      7. ChemACX Ultra DVD Edition (3 above databases on 1 DVD)

      8. ChemINDEX Ultra DVD Edition (4 above databases on 1 DVD)

c. **Cambridgesoft Enterprise Solutions:** Chemistry working environment; Chemistry registration systems http://www.cambridgesoft.com/solutions/

i. ***ChemOffice WebServer*** is an enterprise server for development and deployment of E-Notebook Enterprise, Document Manager, Discovery LIMS, 21CFR11 Compliance, The Merck Index, Registration System, Formulations & Mixtures, Inventory Manager, CombiChem Enterprise, BioAssay HTS, BioSAR Broswer, ChemACX Database and Chemical Databases.

ii. ***Oracle Cartridge*** adds chemical data types to Oracle, linking chemical applications to enterprise software systems.

iii. ***E-Notebook Enterprise*** allows easy compilation of daily research notes. Notebook pages consist of Excel spreadsheets, Word documents, ChemDraw drawings, and spectral data, and can be searched by text and structure.

iv. ***Document Manager*** provides ChemFinder for Word functionality for researchers across the enterprise, allowing searches for chemical structures in documents, files, and folders.

v. ***Discovery LIMS*** features a web interface to initiate laboratory requests, track progress and report results and maintains the full history of a request, including its status and audit trail, and allows requests to be sub-assigned within a lab.

vi. ***21CFR11 Compliance*** is a combination of software (such as E-Notebook and Document Manager) and consulting, analysis, and implementation services designed to assist organizations in complying with government regulations.

vii. ***Chemical Registration System*** is a customizable application that assigns corporate registration numbers for compounds based on each organization's business rules.

viii. ***Formulations & Mixtures*** registers and manages chemical entities that are not limited to pure compounds. This system uses the Chemical Registration application as its back end and adds the capability to find and display all components of a mixture, as well as all formulations that use one or more substances.

ix. ***Inventory Manager*** is a personal-productivity system that uses ChemFinder for Excel and ChemFinder tools for the management of chemical inventories.

x. ***CombiChem Enterprise*** is a reaction-based library generation program that provides the means to integrate library management with subsequent analytical data on an enterprise-wide basis.

xi. ***BioAssay HTS*** allows storage, retrieval, and analysis of biological data used in lead optimization experiments.

xii. ***BioSAR Browser*** is a catalog-driven data mining and analysis application that presents both form and table views within a single web-based interface.

d. **CambridgeSoft Databases:** http://www.cambridgesoft.com/databases/ Key capabilities and offerings of specific databases that can be searched with

ChemDraw or ChemDraw Pluggins:

i. ChemACX Database is a structure-searchable database of 500,000 products from suppliers such as ChemBridge, Maybridge, and Sigma-Aldrich's Rare Chemical Library. Chemical purchasing is accomplished using a shopping cart system works with requisition forms and purchasing system.

ii. The Merck Index is searchable by ChemDraw structure, substructure, names, partial names, synonyms and other data fields by desktop, enterprise and online formats.

iii. Organic Syntheses provides detailed experimental methods in a standard format for the synthesis of organic compounds. In 79 annual volumes and 9 collective volumes, *OS* compiles independently checked selected procedures and new reactions which permit the advanced research student or professional chemist skilled in the field to prepare compounds of research utility.

iv. ChemINDEX provides access to the four databases described in Section IIIa, subsection ix.

v. World Drug Index (WDI) from Derwent contains over 58,000 compounds with known biological activity. WDI classifies compounds according to type of biological activity, mechanism, synonyms, trade names, and references.

vi. Current Chemical Reactions (CCR) from ISI contains information from over 300,000 articles reporting the complete synthesis of molecules updated daily.

vii. ChemReact and ChemSynth allow synthetic chemists to design novel syntheses from information carefully selected from a database based on over 2.5 million reactions.

viii. World Drug Alerts (WDA) from Derwent is a current awareness application providing information on patents, new biologically active compounds, new methods for synthesizing drugs, and other data.

ix. Investigational Drugs Database (ID db) from Current Drugs is a competitor intelligence service on drug R&D. Updated weekly, it covers aspects of drug development world wide, from first patent to launch or discontinuation.

IV. Review:

ChemDraw Ultra 9.0 is a versatile drawing package with a very intuitive GUI that converts chemical structures to chemical names and vice versa and generates both proton and carbon NMR spectra from structures. ChemDraw ActiveX/Plugin Pro 9.0 allows the interaction with online databases to query, mine and view data, and it can also be used to publish online structures. CambridgeSoft's complete Desktop and Enterprise product line is described in a 28-page brochure: http://products.cambridgesoft.com/datasheets.cfm

- *Desktop Software*: ChemOffice, E-Notebook, ChemDraw, Chem3D, ChemFinder, ChemInfo

- *Enterprise Solutions*: ChemOffice WebServer, Oracle Cartridge
- *Knowledge Management*: E-Notebook Enterprise, Document Manager, Discovery LIMS, 21CFR11 Compliance
- *Research & Discovery*: ChemOffice WebServer, Oracle Cartridge
- *Applied Bioinformatics*: BioAssay HTS, BioSAR Browser
- *Chemical Databases*: ChemACX Database, ChemSAR Properties, The Merck Index, Chemical Databases

CambridgeSoft's collaborations with companies such as InfoChem (for databases) and Spotfire (for data visualization) allow chemists to access a wide range of chemistry and reaction information using CambridgeSoft's tools. In addition, CambridgeSoft also provides access to other sources of information such as commercially available compounds, screening compounds, reaction databases, and patent information. Chem3D provides a powerful platform to carry out molecular calculations and compound visualization. In this regard it is noted that Chem3D does not offer a batch-mode feature to minimize a set of molecules using empirical or semi-empirical or semi-empirical calculations. Capability to dock small molecules against large molecules (i.e., proteins) is not supported at the present time.

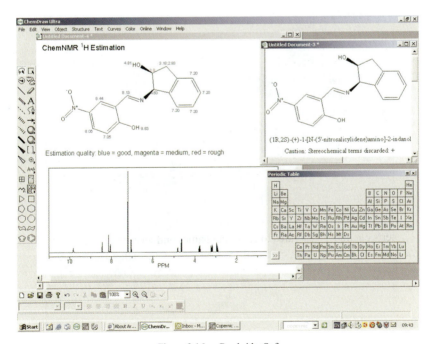

Figure 2.16. CambridgeSoft

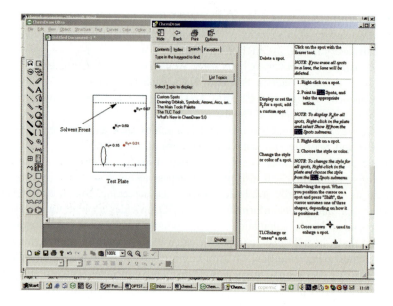

Figure 2.17.   CambridgeSoft

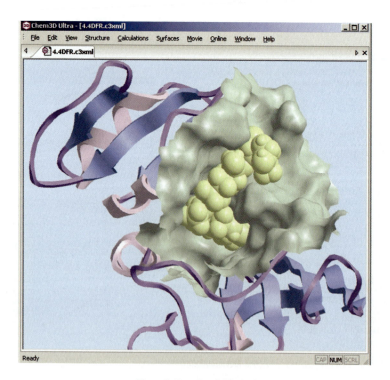

Figure 2.18.   CambridgeSoft

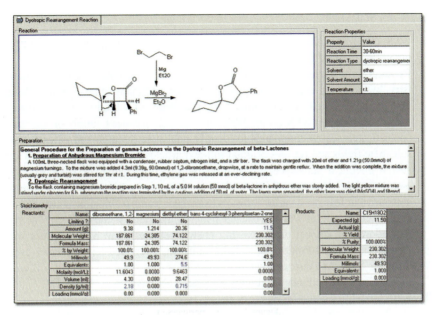

Figure 2.19.   CambridgeSoft

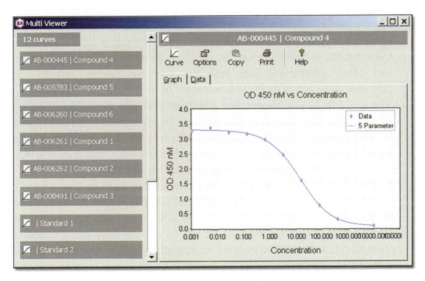

Figure 2.20.   CambridgeSoft

2.6   CAS/Scifinder

   I. CAS, Chemistry Abstract Services, www.cas.org, http://www.cas.org, and http://
      www.cas.org.prod.html
  II. **Product Summaries:** STN contains data from over 200 databases covering chem-
      istry, life sciences, engineering, patents, physics, and many other scientific fields
      that can be searched using command line searching, STN Express with Discover!
      using "wizards," or STN on the Web. One can ask questions simply or use more
      sophisticated search commands to identify published research and patents in all
      scientific fields and retrieve original full-text articles and patents on the Web and
      search chemical substance information by structure, name, or CAS Registry
      Number (CAS Number). STN AnaVist is interactive analysis and visualization
      software that features a variety of ways to analyze search results from scientific lit-
      erature and patents as well as visualize patterns and trends in the research. The sys-
      tem allows the import of an existing answer set created in STN Express with
      Discover!, Analysis Edition, Version 8.0, or use the integrated concept search capa-
      bility in STN AnaVist. Similar information can be obtained through different
      mechanisms via SciFinder SciFinder Scholar, STN Easy, and CA on CD.
 III. Key capabilities and offerings:
      a. **Over 200 databases are accessible through STN**, for a full listing with the
         details see: http://www.cas.org/ONLINE/DBSS/dbsslist.html. STN uses CA
         Lexicon for searching and permits structure-based searches similar to those
         available through SciFinder. STN Database Summary sheets are produced for
         every file on STN. Each sheet describes file content, sources of the file, file
         data, and producer. Note: Japanese language versions of STN database sum-
         mary sheets are available as PDF files from JST.
      b. **SciFinder** provides access to more than 23 million abstracts, over 26 million
         organic and inorganic substances, and over 56 million sequences. SciFinder
         is one of several interfaces for access to the voluminous information in the
         CAS databases. SciFinder can be searched by structure, concepts, reaction
         and other parameters and searches may be further refined. See:
         http://www.cas.org/SCIFINDER/scicover2.html
      c. **The CA File is a bibliographic database** available from CAS (Chemical
         Abstracts Service) covering international journals, patents, patent families,
         technical reports, books, conference proceedings, and dissertations from
         areas of chemistry, biochemistry, chemical engineering, and related sciences
         from 1907 to the present, with some records back to 1840.
      d. **CAS REGISTRY** is a substance database containing more than 26 million
         organic and inorganic substances and 56 million sequences. CAS Registry cov-
         ers all substances indexed in CA from the mid 1950s to the present, and sub-
         stances from the early 1900s to the 1950s are currently being added. All
         substance records contain a unique CAS Registry Number and may also have CA
         index names, synonyms, molecular formulas, physical and chemical properties,
         and structure diagrams, all of which are searchable and displayable. Updated

daily, the database covers substances identified from the scientific literature as well as from registrations from regulatory lists such as TSCA and EINECS.

e. **Each CAS Registry Number is a unique numeric identifier**, designates only one substance and has no chemical significance. It provides a link to a wealth of information about a specific chemical substance. A CAS Registry Number is assigned in sequential order to each new substance when it enters the CAS REGISTRY database.

f. **CA on CD**: CA on CD, which is available on an annual subscription basis, provides more than 814,000 references from more than 9,500 major journal articles and patents from over 50 active patent-issuing organizations around the world. *Chemical Abstracts* (CA) is divided into 80 different sections. See: http://www.cas.org/PRINTED/sects.html for a full listing.

g. **CAplus contains the most current and most comprehensive chemistry bibliographic database available from CAS** (Chemical Abstracts Service). CAplus is available on *STN International*, *STN Easy*, *STN on the Web*, *SciFinder*, and *SciFinder Scholar*. CAplus contains more than 23,000,000 records (08/03) from 1907 to the present. Automatic current-awareness searches (SDI's) may be run daily, weekly (default), or biweekly. More than 9,000 journals monitored and patents (and patent families) from some 50 patent-granting organization around the world as well as cover-to-cover coverage for nearly *1,500 key chemical journals* (since October 1994) including records for huge variety of scholarly document types not covered in Chemical Abstracts CA. For more information on each of these categories see: http://www.cas.org/chemplus/chemplus1.html#material.

h. **MARPAT – the CAS Markush database contains the keys to prophetic and generic substances in patents**. MARPAT is a structure-searchable database of generic and hypothetical substances found in patent claims and disclosures. MARPAT is available on *STN* and *STN on the Web*. MARPAT is part of the CASLINK cluster on STN. Through CASLINK structure searches are performed not only in MARPAT but also MARPATprev and REGISTRY. Answers are documents with information on the substances containing the query structure. More than 582,000 (7/03) searchable Markush structures from patents are covered by CAS from 1988 to the present with more than 214,000 (7/05) displayable citations for the Markush structure-containing patents with displayable information includes bibliographic data, abstract, and CAS indexing from 44 patent-granting organizations.

i. **CASREACT provides access to information on synthetic organic research**, including organometallics, natural products, and biocatalyzed reactions it is searchable by reaction, product, reactant or reagent information concerning reaction conditions, yields and catalysts. CASREACT contains content from 1840 to the present with more than 9 million single- and multi-step reactions from more than 501,000 records from journal articles and patents with reaction information and has automatic current-awareness searches (SDIs) run every week.

j. **CHEMCATS provides information on commercially available chemicals and their worldwide suppliers** including more than 8.7 million records and 9.2 million products from 722 suppliers and 901 catalogs (updated 7/05). Each record contains the catalog information for the substance e.g., chemical and trade names, the company names and addresses, as well as supplier information, and, pricing terms. More information is available at: http://www.cas.org/ONLINE/DBSS/chemcatsss.html.

IV. Review:

SciFinder provides an easy interface to search efficiently scientific information without needing to learn complicated issues of database searching of chemical information. Tutorials are available on the web for using Explore by Research Topic, how to set up a Keep Me Posted alert to get breaking news, Browse the Table of Contents of journals, and other SciFinder features. Explore by Chemical Structure allows one to find substances based on their structure to display its physical properties, as well as information on obtaining the substance from commercial sources. CAS' STN and SciFinder are considered the most extensive source of chemical information, particularly for information from the patent literature.

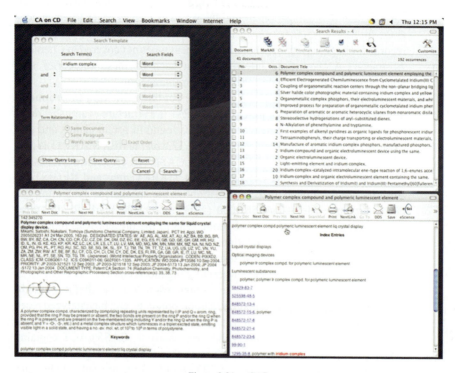

Figure 2.21.   CAS

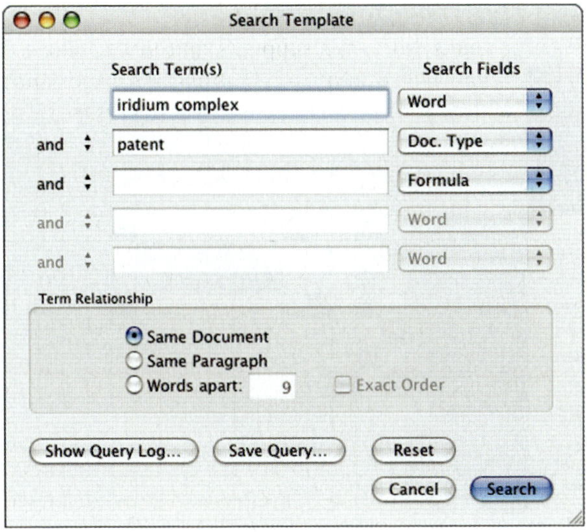

Figure 2.22. CAS

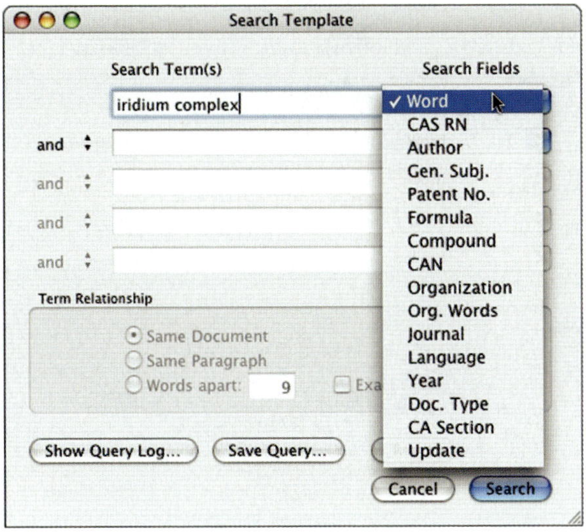

Figure 2.23. CAS

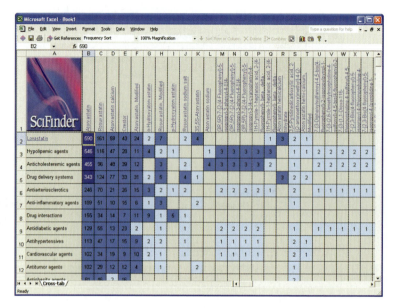

Figure 2.24.   CAS

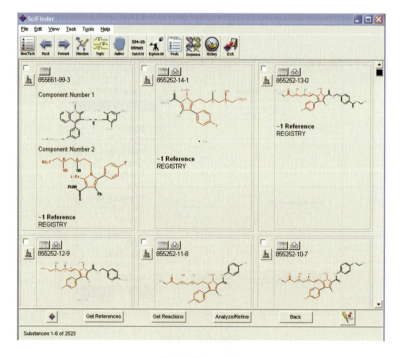

Figure 2.25.   CAS

Figure 2.26.   CAS

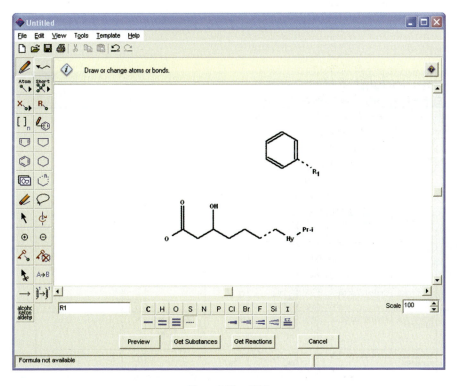

Figure 2.27.   CAS

A (Furanone)	References	BIOL (Biological study)	USES (Uses)	THU (Therapeutic use)	BSU (Biological study, unclassified)	PAC (Pharmacological activity)	PREP (Preparation)	SPN (Synthetic preparation)	RCT (Reactant)	RACT (Reactant or reagent)	ADV (Adverse effect, including toxicity)	PRP (Properties)	BAC (Biological activity or effector, except adverse)	PROC (Process)	BPR (Biological process)	DMA (Drug mechanism of action)	ANST (Analytical study)	PEP (Physical, engineering or chemical process)	PKT (Pharmacokinetics)	PYP (Physical process)	ANT (Analyte)	DGN (Diagnostic use)	FFD (Food or feed use)	BUU (Biological use, unclassified)
8696 162011-86-1	6	5	5	5	4	1	6	6					4											
8697 162011-87-2	5	4	4	4	4		5	5					4											
8698 162011-88-3	5	4	4	4	4		5	5					4											
8699 162011-89-4	7	4	4	4	4		6	6	2	2			4											
8700 162011-90-7D	31	30	28	28	8	18	4	4				4	7	1				1		1				
8701 162011-90-7	1063	1008	944	932	151	526	26	24	14	14	241	39	113	30	7	61	38	23	22	17	38	2	7	7
8702 162011-91-8	5	4	4	4	4		5	5					4											
8703 162011-92-9	5	4	4	4	4		5	5					4											
8704 162011-93-0	5	4	4	4	4		5	5					4											

Cross-tab / 3-D Column

Figure 2.28.   CAS

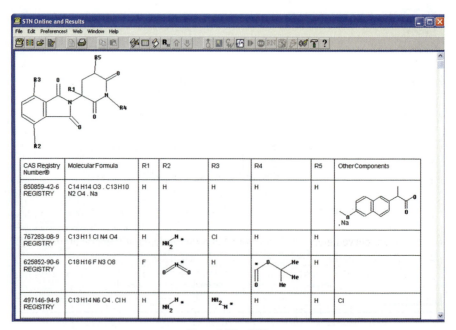

CAS Registry Number®	Molecular Formula	R1	R2	R3	R4	R5	Other Components
850859-42-6 REGISTRY	C14 H14 O3 . C13 H10 N2 O4 . Na	H	H	H	H	H	(structure), Na
767283-08-9 REGISTRY	C13 H11 Cl N4 O4	H	NH2 (structure)	Cl	H	H	
625852-90-6 REGISTRY	C18 H16 F N3 O8	F	(structure)	H	(structure)	H	
497146-94-8 REGISTRY	C13 H14 N6 O4 . Cl H	H	NH2 (structure)	NH2 N*	H	H	Cl

Figure 2.29.   CAS

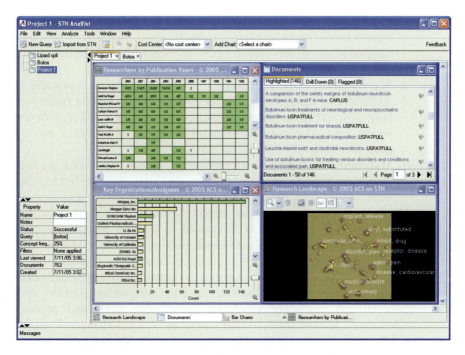

Figure 2.30. CAS

2.7 CHEMAXON

I. ChemAxon; http://www.chemaxon.com/

II. Product Summary

    a. **General:** ChemAxon provides Java based chemical software development platforms for the biotechnology and pharmaceutical industries. ChemAxon creates web-based cross platform solutions for chemoinformatics and chemical communication.

    b. *Marvin* is a collection of Java tools for drawing, displaying and characterizing chemical structures, substructures and reactions. It features advanced molecule (isotopes, radicals, lone pairs, templates, abbreviated groups, multiple groups, attached data, 3D sketching), reaction (manual and automatic mapping, reaction stereo, curved electron flow arrows) and query drawing (generic atoms/bonds, atom lists/notlists, pseudo atoms, link nodes, other query properties, recursive SMARTS, R-groups) capabilities. Marvin contains 2D cleaner for presentation quality molecule display and a proprietary geometry optimizer able to create 3D geometry from connectivity and perform conformational search.

        i. **Marvin Applets:** Tools for building chemical web pages, which are compatible with most browsers (Internet Explorer, Netscape, Mozilla, Firefox,

Safari, Opera, etc.) and have two GUIs: AWT and Swing. They offer access from/to JavaScript and are customizable by applet parameters. The signed versions of the applets support access to local files (open/save molecule files, save structure images).

ii. **Marvin Beans:** A set of classes for building applications. It provides an API with low and high level classes, and JavaBeans. Marvin Beans support copy and paste to/from several other structure handling software, import/export various formats (SMILES, Molfile, SDfile, etc.), and can generate images (BMP, PNG, JPG, SVG, etc.)

iii. **MarvinSketch/MarvinView:** Applications for end users which are built from Marvin Beans. MarvinSketch is a tool for drawing molecules/reactions. MarvinView displays a set of structures.

c. *MarvinSpace* is an OpenGL based high performance 3D molecule visualization tool written in Java. It is WEB enabled, its seamless integration in html pages makes it an ideal platform neutral molecule visualizer. MarvinSpace can visualize small molecules, proteins, nucleic acids, crystals, various molecular surfaces, volumetric data such as electrostatic potential, hydrophobicity. It provides automated methods for molecular overlay, geometry optimization and conformer generation using force fields and property calculations. Manual overlay of structures and manual change of internal coordinates are also available. Beyond visualization MarvinSpace allows to share molecular information by labeling and annotating atoms, bonds, ligands, receptors, complexes, surface regions.

d. *Calculator Plugins* are an open technology, custom chemical calculation platform for Marvin and other JChem tools. Default calculations are provided as dynamically loading- plugins. Currently available tools from ChemAxon include:

i. pKa
ii. log P, log D
iii. polar surface area (PSA)
iv. charge distribution
v. Hückel analysis
vi. polarizability prediction
vii. H-bond acceptors/donors
viii. major microspecies
ix. refractivity
x. isoelectric point
xi. tautomers
xii. resonance
xiii. elemental analysis
xiv. topology analysis
xv. http://www.chemaxon.com/marvin/chemaxon/marvin/help/calculator-plugins.html provides the theory behind the programs used for the calculations.

e. *JChem Base* is a Java tool for the development of applications that allow for the search of mixed structural and non-structural data. JChem Base can also integrate a variety of database systems (Oracle, MS SQL Server, DB2, Access, MySQL and PostgreSQL) with web interfaces and offers fast substructure, similarity, and exact search engine using 2D hashed fingerprints. Structures are stored in database tables. Structural and non-structural data can be combined. SDF, SMILES, etc. can be imported and exported. The system includes Marvin, a Java based chemical editor and viewer for the following:

   i. Graph based topology search algorithm

   ii. Hashed chemical fingerprint-based database search accelerator technology

   iii. Pharmacophore fingerprints

   iv. Custom structural keys

   v. Exact structure, substructure, superstructure and similarity matching, generic query atoms and bonds, atom lists and notlists, SMARTS and query property support, recursive SMARTS, R-group queries, reaction searching (component level grouping, inversion/retention flags), stereosearching (chirality centers, double bonds, diastereomers with enhanced stereo attributes). See query features in details at http://www.chemaxon.com/jchem/doc/user/Query.html.

   vi. The chemical terms language built by ChemAxon combines structure based calculation functions with structure queries. Currently more than a hundred functions are available through the plugin system providing a new level of flexibility for defining chemical queries. It is simple to formulate Lipinski's rule of five, lead-likeness, bioavailibility or other complex filters in chemical terms. (Chemical Terms can also be used for pharmacophore definitions, drug design goal functions or for defining reaction rules).

f. *JChem Cartridge* Using the JChem Cartridge for Oracle the non Java developers (VB, C++, etc.) can also access many JChem functions, such as structure searching or property predictions, are available from within Oracle's SQL.

g. *Standardizer* is a structure canonization tool in JChem for converting molecules from different sources into standard representational forms. Standardizer can automate the identification of mesomers and tautomers, can be used for counter-ion removal, and offers many more functions for preparing compounds for database storage. It provides a flexible molecule transformation engine for:

   i. hydrogen manipulations

   ii. aromatic bond transformations

   iii. mesomers

   iv. tautomers

   v. custom functional group transformations

   vi. solvent and counter ion removal

   vii. ungrouping Sgroups

      viii. removing stereo features

       ix. removing or setting the chiral flag

        x. expanding by stoichiometry

       xi. cleaning 2D coordinates

      xii. performing template based cleaning

h. *Screen* is a HTS suite in JChem, which works with files and structure data-bases and features various models for the similarity analysis of molecules and pharmacophore hypotheses. The pharmacophore mapping tool offers cus-tomizable pharmacophore models as well as an optimizer to find the "best" screening metrics and parameter sets for screening compounds similar to a given active compound family.

        i. configurable molecule descriptor generation

          1. hashed chemical fingerprint;

          2. pharmacophore fingerprint

          3. BCUT

          4. scalar descriptors

          5. user defined descriptors

       ii. pharmacophore mapping

      iii. hypothesis fingerprints

      iv. virtual screening;

          1. Tanimoto and Euclidean metrics

          2. weighted, scaled, normalized and directed

       v. Screening Optimization

i. *JKlustor* is a tool of JChem for clustering, diversity calculations, and library comparisons based on molecular fingerprints and other descriptors. JKlustor is useful in combinatorial chemistry, drug design, or other areas where a large number of compounds need to be analyzed. Beside the fingerprint based Ward and Jarvis Patrick methods, JKlustor provides a maximum common substructure (MCS) based clustering tool able to visualize the common struc-tural elements of the compound hierarchy.

j. *Reactor* is the virtual reaction engine. It supports "smart" reactions (generic reaction equations combined with reaction rules) generating synthetically feasible products even in batch mode. The professional version of the tool includes support for multi-step virtual synthesis and filtering of chemically feasible molecules which are not of interest. Reactor is an effective tool for synthesizing in silico combichem libraries, simulate metabolic biodegrada-tions or for building diverse compound spaces of synthesizable molecules. A free synthesis library is also provided by ChemAxon.

k. *Fragmenter* creates building blocks by fragmenting larger molecules. It comes with a built in RECAP module, but the fragmentation rules are customizable. The original connection information are stored with the fragments. Fragmenter is particularly relevant for generating analogues of biologically active compounds for lead discovery. When activity information is provided for a compound library, an additional statistics tool can be used to determine key structural elements responsible for high or low activity. Fragmenter is shipped with an R-group decomposition tool able to identify the ligands connected to a common scaffold of a combichem compound library.

III. Key capabilities and offerings:

   a. Marvin is a Java based chemistry software that is available in various forms. **Marvin Applets** are created for the web developer who builds chemical Internet/Intranet sites. Marvin can handle molecules in various *file formats* including MDL mol, Compressed mol, unique SMILES, SMARTS, Sybyl mol, PDB, CML, XYZ, POV-Ray. MarvinView can display a 2D or 3D molecule, or many molecules in a table. Marvin can be equipped with custom chemical calculation tools by the integrated plugin services.

   b. **Marvin** contains a framework for integrating chemical computations into the drawing/viewing application environment. These tools – called plugins – are loaded dynamically upon request. ChemAxon provides various tools for calculating charge, pK_a, log P, etc. The available calculator plugins are located in the **Tools** menu. The corresponding calculation parameters can be set in the parameter panel accessible from the **Options** submenu. Some plugins (charge, polarizability, polar surface area and hydrogen bond donor-acceptor) optionally perform their computation on the *physiological microspecies* of the input molecule taken at a specified pH. This option together with the corresponding pH can also be set in the parameter panel.

   c. **JChem** is a development tool written in Java for manipulating mixed chemical and corporate data for web-based applications. Java supports most operating systems, database handlers, and web servers. JChem is a tool for developing distributed custom chemical applications that can be accessed by web browsers.

      i. *JChem Base* provide a chemical interface to databases, which can be applied for combined SQL and structural queries; imports/exports molecules, substructures, or reactions in standard formats (Molfile, SD file, RD file, SMILES, SMARTS, etc.).

      ii. *JChem Cartridge* provides a chemical cartridge for the Oracle platform giving automatic access to Oracle's security, scalability, and replication features.

      iii. *Standardizer* structure canonization tool converting molecules from different formats into standard representation.

      iv. *Screen* screening based on pharmacophore or chemical fingerprints or other descriptors.

      v. *Reactor* generating reaction products from reaction equations and reactants.

      vi. *Fragmenter* generating building blocks based on Recap rules from molecule libraries.

      vii. *Serial Molecule Generator* transforming molecules by a sequence of user-defined transformations.

      viii. *Chemical Term Evaluator* evaluating chemical expressions.

    ix. *JKlustor* clustering and diversity calculations based on molecular fingerprints or other properties.

d. **Standardization:** the goal of any standardization method is to bring the molecules to standardized form and then perform the search. First for or each alterable functional group (e.g. oxo-enol tautomers) and drawing modes (e.g. aromatic ring – alternated single and double bonds) the standardization program must decide which one to choose. This decision is given in the *Configuration File*. The standardizer performs the necessary transformations on the molecule in the order they are given in the configuration file.

    i. **Input:** Most molecular file formats are accepted (for instance MDL molfile, Compressed molfile, SDfile, Compressed SDfile, SMILES). The input is either specified in input file(s), or else in input string(s), usually in SMILES format. If neither the input file name(s) nor the input string(s) are specified in the command line then the standard input is read.

    ii. **Output:** Standardizer writes output molecules in the format specified by the – format option (the default format is "smiles"). If the – output is omitted, results are written to the standard output.

e. **Screen:** The Screen software suite provides tools for pharmacophore analysis, screening and selectivity optimizations. Biologically active compounds usually bind to their specific receptors by weak forces. These molecular interactions include hydrogen binding, electrostatic attraction, hydrophobic affinity, etc. The existence of so-called *pharmacophore points* in the appropriate positions is often required for receptor binding. The pharmacophore pattern of a molecule can be determined with the help of the *Pharmacophore Mapper* utility. The comparison of molecular structures is based on the comparison of *molecular descriptors*. The analysis of the *chemical* and *pharmacophore* topology using fingerprints helps recognize molecules having similar biological activities. The *ScreenMD* utility can select molecules from huge molecule databases similar to biologically active compounds such as drugs. Sometimes the structure of the found molecules is not chemically similar to known active compounds, but they are similar to the pharmacophore hypothesis. ChemAxon developed a *software package* that helps to find the best parameter set (descriptors, metrics, normalization, weighting, scaling, directing) for each compound family. The *enrichment ratio* of the screening procedure can be higher using an optimized parameter set compared to unoptimized standard screening methodologies.

f. **JKlustor** is a Java software for diversity calculations and clustering. The module is integrated with JChem (its classes are included in jchem.jar). Though the target audience of the system are chemists, it can be used for any types of objects, not just molecules. At present, JKlustor includes the following command-line tools:

    i. *GenerateMD* generates various molecular descriptors including chemical hashed fingerprints and pharmacophore fingerprints for molecules, which may be used for structural diversity computations and clustering.

   ii. *Compr* compares two sets of objects (like compound libraries) using diversity and dissimilarity calculations.

g. **Reactor** simulates a chemical reaction: given the reaction and the reactants, the program creates the products that the reactants would be transformed to by the given chemical reaction. The reaction defines the way that the reactants will be converted to products: determines the bonds/atoms that are removed, the bonds/atoms that are created and those that are transformed (e.g. from single to double bond). This is done with the help of *atom maps*. Atom maps should be unique on both the reactant side and the product side, bonds are changed where both ending atoms are mapped. Atoms on the reactant side with atom maps not existing on the product side are removed, atoms on the product side with atom maps not existing on the reactant side are created, bonds connecting mapped atoms are copied from the product side while others are copied from the reactant side. Reactor supports the following reaction file formats: RXN, RDF, SMIRKS. When using SMIRKS, all features of *reaction SMARTS* are available, without the limitations of SMIRKS.

h. **Fragmenter** fragments molecules based on predefined cleavage rules. The cleavage rules are given in form of reaction molecules in the *configuration XML*. By default, all non-ring bonds matching the cleavage bonds in the rules are cleaved. However, it is possible to provide a revision algorithm that forbids certain cuts depending on predefined criteria (e.g. the resulting fragment size, the structural environment of the bond, the number of cleaved bonds in the resulting fragments, etc.). Currently one such algorithm is implemented: the RECAP method. The RECAP algorithm following a number of cleavage revision rules (never cut a hydrogen-connecting bond, never cut a bond connecting a ring-carbon and a hetero atom, etc.).

IV. Review:

ChemAxon provides platform independent, java-based chemoinformatics tools. Drawing packages are available through the internet for chemistry end-users. Web developer and software developer tools allow programmers to customize their own views and features. Marvin allows structure drawing and viewing. The calculator plugins facilitate standard structure characterization (MW, cLog P, etc.). A review of the theory behind the various computational methods is provided: http://www.chemaxon.com/marvin/chemaxon/marvin/help/ calculator-plugins.html#ms Taken together they allow the communication of chemical and non chemical data over networks without client software installation. JChem Base allows structure searching and database access from a JChem Cartridge which integrates with an Oracle database. ChemAxon provides FREE academic license for ALL of its products (see more details in ChemAxon's forum at http://www.chemaxon.com/forum/ftopic193.html).

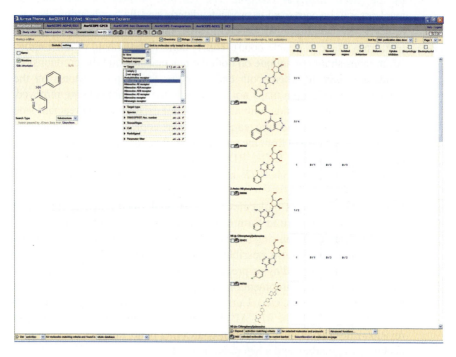

Figure 2.31.   This screenshot shows the Aureus Pharma product AurQUEST where Marvin, MarvinView
and JCHEM from ChemAxon are utilized. Using AurQUEST, a researcher can query Aureus Pharma's
AurSCOPE knowledgebases to gain both chemical & biological insights to help aid in accelerating drug
discovery. Using Marvin, a user can sketch a chemical compound of interest and search using JCHEM's
substructure or other chemical searching methods to find compounds of interest. In this example the
chemical searching has been combined with a biological query in which we are searching for ligands that
have been tested to bind the G-Protein Couple Receptor target Adenosine A1. The results of the query are
visualized using MarvinView on the right handside of the graphic

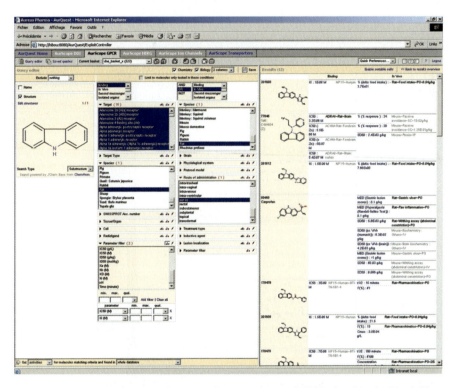

Figure 2.32. This screenshot shows the Aureus Pharma product AurQUEST where Marvin, MarvinView and JCHEM from ChemAxon are utilized. Using AurQUEST, a researcher can query Aureus Pharma's AurSCOPE knowledgebases to gain both chemical & biological insights to help aid in accelerating drug discovery. Using Marvin, a user can sketch a chemical compound of interest and search using JCHEM's substructure or other chemical searching methods to find compounds of interest. In this example the chemical searching has been combined with a biological query in which we are searching for ligands that have been tested to bind to a variety of G-Protein Couple Receptor targets and have also been tested in vivo experiments. The results of the query are visualized using MarvinView on the right handside of the graphic

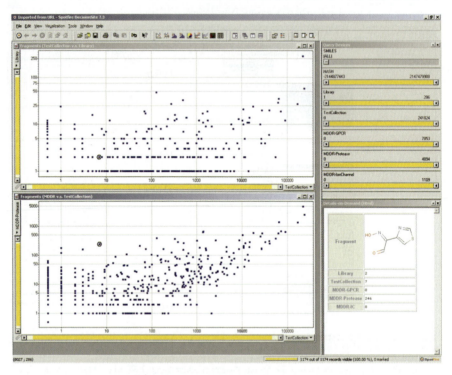

Figure 2.33.   A fragment analysis application developed by AstraZeneca by integrating Fragmenter and Marvin in Spotfire. The application provides integrated views on hit frequencies of the fragments from a library in various targeted compound collection spaces. The structure and hit frequencies for the fragments are displayed in the lower-right corner while the user can interactively click one fragment dot or select a set of fragments. This screenshot is provided by Paul Xie of AstraZeneca

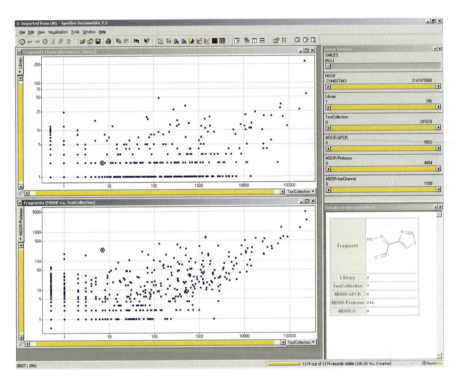

Figure 2.34. Ligand, methotrexate bound to the protein dihydrofolate reductase visualized in MarvinSpace. Detailed view of the binding pocket. The surface is colored by residue type and it is transparent. Behind it the ball and stick representation of the protein is seen. Lengths of some hydrogen bonds are monitored. Two pharmacophore regions are also defined; an acceptor (red sphere) and a hydrophobic region (large yellow sphere), these are also semi-transparent. Image courtesy of ChemAxon

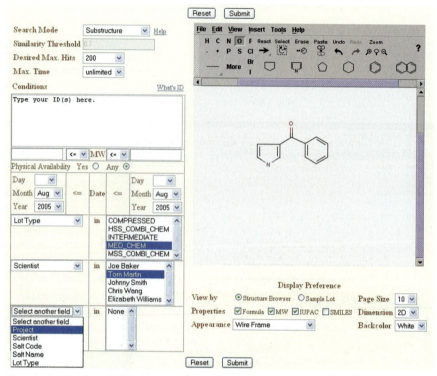

Figure 2.35.   Web-based chemistry search tool that combines multiple search criteria to query compound structures and related data. Uses Marvin for structure input and viewing and JChem Base for structural search, designed by Zhenbin Li of Neurogen Corporation

2.8   CHEMICAL COMPUTING GROUP

I. Chemical Computing Group (http://www.chemcomp.com/)

II. Product Summary: The Chemical Computing Group (CCG) offers its MOE (Molecular Operating Environment) computational software platform for life science applications such as Bioinformatics, Chemoinformatics, High-Throughput Discovery, Structure-Based Design, Protein Modeling, Molecular Modeling and Simulations, and Methodology Development and Deployment.

III. Key capabilities and offerings:

   a. *Bioinformatics*: This set of tools allows the access to protein crystallography data for structure analysis, protein structure determinations from amino acid sequences, and protein structure predictions.

      i. Protein Structure Database. This database is a curated version of the Protein Data Bank containing searchable fields such as Code, Header, Compound, Title, HET groups, resolution, etc.

      ii. Structural Family Database. GCC has processed and clustered the proteins of the Protein Data Bank to provide a database of structural families.

    iii. Fold Identification. A fold detection algorithm allows for the searching and identification of structurally similar proteins in the Structural Family Database.

    iv. Multiple Alignment. The multiple protein alignment methodology implemented in MOE can correlate the sequence and structures of large collections of proteins.

    v. Structural Family Analysis. Allows the study and correlation of conserved features and differences between related protein structures and homologous sequences.

b. *Chemoinformatics*: This suite allows the manipulation and analysis of large sets of chemical structures.

    i. Molecular Databases. These spreadsheet-based databases can store and manipulate large number of compounds. The structures can be imported through several formats and structures can be automatically stripped from salts and solvents.

    ii. Molecular Descriptors. This system can calculate 400 types of molecular descriptors (e.g., topological indices, structural keys, physical properties, topological polar surface area) and use custom-made descriptors using MOE's built-in Scientific Vector Language.

    iii. QSAR/QSPR Predictive Modeling. This set of tools permits the construction of models to predict activity, cluster and filter collections of compounds, and carry out diversity and similarity assessments.

    iv. Molecular Fingerprinting / Clustering / Similarity Searching. 2D and 3D molecular fingerprinting systems including MACCS, 3D shape, and pharmacophore fingerprints for similarity searching, clustering and diversity analysis.

    v. High Throughput Conformational Search. Keeping a database of molecular fragments, the system is capable of executing fast conformational searches on large collection of compounds.

    vi. 3D Pharmacophore Search. The system allows the construction of queries of pharmacophoric features working in sync with the conformational database mentioned above.

    vii. 3D Pharmacophore Elucidation. An application for automatically generating pharmacophore queries based on acitivity data.

    viii. Suite of SDF file processing utilities for compound database management.

    ix. 2D depiction tools for reports and presentations.

c. *High-Throughput Discovery*: MOE comes with an integrated set of tools for the handling and process of combinatorial compound libraries coming from combinatorial chemistry and high-throughput screening.

    i. Molecular Descriptors. Same as described in CCG's Chemoinformatics section (b).

    ii. HTS-QSAR. All molecular structures are processed and structure-property correlations are determined using linear or probabilistic methodologies.

   iii. Combinatorial Library Enumeration. MOE comes with an integrated enumerating engine to construct combinatorial compound libraries with chirality awareness.

   iv. Focused Combinatorial Library Design. Focused combinatorial libraries are designed by incorporating QSAR and ADME models to propose compounds with a higher probability of desired biological activity. The compounds are enumerated using a product-based approach.

   v. Diverse Combinatorial Library Design. Large sets of diverse combinatorial libraries can be product-based enumerated using a Monte Carlo sampling technique to extract diverse compound subsets.

d. *Structure-Based Design*: Using crystallographic data of macromolecules MOE can assist in the study, visualization and identification of active sites and receptor-ligand interactions to design and screen new ligand candidates.

   i. Active Site Detection. Using a geometric algorithm, MOE can detect candidate protein-ligand and protein-protein binding sites.

   ii. Probabilistic Contact Potentials. The system maps and identifies hydrophobic and hydrophilic contact sites between target and ligand determining inter-atomic distances and angles.

   iii. Multi-Fragment Search. The program populates a large number of chemical fragments in a predetermined active site where the fragments are minimized, clustered and scored for their analysis.

   iv. Ligand-Receptor Docking. Module-based docking protocol for placing ligands in biding sites. Docking modules are extensible allowing for easy incorporation of different placement algorithms and scoring methods. Solvation effects can be included in the calculations.

   v. 3D Pharmacophore Elucidation. An application for automatically generating pharmacophore queries based on activity data.

   vi. 3D Pharmacophore Search. The system allows the construction of queries of pharmacophoric features working in sync with the embedded conformational database to identify most promising ligands.

e. *Protein Modeling*: MOE comes with an integrated protein tool set for protein structure prediction.

   i. Homology Search. The system searches a database of structural families calculated by 3D clustering the Protein Data Bank for sequence-to-structure predictions.

   ii. Multiple Alignment. The multiple protein alignment methodology implemented in MOE can correlate the sequence and structures of large collections of proteins.

   iii. Structural Family Analysis. Allows the study and correlation of conserved features and differences between related protein structures and homologous sequences.

   iv. Structure Prediction. The system can take an amino-acid sequence input, match the sequence against experimentally determined backbone structures, and refines the predicted structure using the AMBER '89/'94/'99, CHARMM22/27 or Engh-Huber forcefields.

    v. Structural Quality Assessment. To assess the reliability of predicted structures MOE uses statistical measures derived from X-ray crystallographic data. Diagnostic measures include Ramachandran and Chi plots.

f. *Molecular Modeling and Simulations*: In order to construct, study and model chemical structures MOE has embedded molecular editors and validated forcefields.

    i. Molecular Builders and Data Import/Export. MOE comes with integrated building interfaces to create or edit small molecules, proteins, carbohydrates, DNA and crystal structures. The system can also import and export structures in many standard file formats.

    ii. Molecular Mechanics. MOE's engine integrates and uses different forcefields such as AMBER '89/'94/'99, CHARMM22/27, MMFF94, MMFF94s, OPLS-AA and Engh-Huber.

    iii. Implicit Solvent Electrostatics. MOE uses a multi-grid algorithm to study and predict solvation effects without explicit treatment of water molecules.

    iv. Conformational Analysis. The system can perform conformational analysis using Molecular Dynamics, Hybrid Monte Carlo, Stochastic or Systematic search methodologies.

    v. Flexible Alignment of Small Molecules. The program uses an all-atom flexible alignment procedure to align or superimpose 3D structures.

    vi. Diffraction Simulation. MOE can simulate X-Ray, Neutron or Electron diffraction experiments in either gas phase, amorphous liquid, powder and single crystal phases.

    vii. Quantum Calculations. Ability to launch QM jobs such as MOPAC, GAMESS and Gaussian from MOE.

g. *Methodology Development and Deployment*: MOE's design allows the integration of other applications or the creation of new ones for Life Sciences.

    i. Scientific Vector Language (SVL). SVL is a high-level scripting language to develop applications with MOE.

    ii. Background Computing. MOE/batch can run batch or background calculations that do not require a graphical interface.

    iii. Cluster Computing. MOE/smp allows the use of multiple computers to perform and carry out large-scale calculations.

    iv. Computer Platforms. MOE can be used with Intel computers running Microsoft Windows or Linux as well as IBM eServer, Sun Microsystems, Hewlett-Packard, MOE Os X and Silicon Graphics computers running Unix.

IV. Review: MOE integrates under one platform tool sets for the study of macromolecules and the design of promising inhibitors using combinatorial chemistry and receptor-ligand information with ADME and QSAR models. The MOE system can operate under a wide range of computer platforms to minimize the IT barrier. The capability of running calculations in batch mode is a useful option to maximize the use of the system during non-peak hours (i.e., nights, weekends, etc.). Scalability is another feature of the system since calculations can be carried out using a computer farm. The system requires the user learn the SVL language to develop applications, something that medicinal and combinatorial chemists are not always receptive to do. However, the language is efficient and intuitive to learn and CCG provides customized training to commercial users.

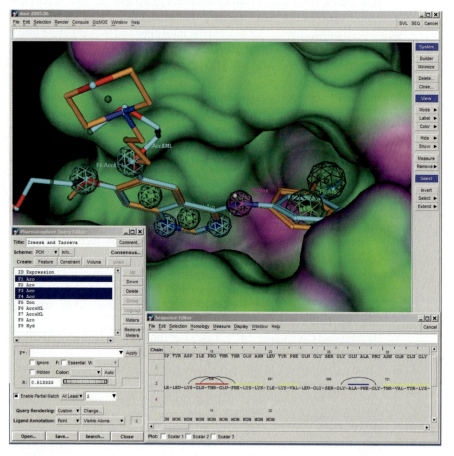

Figure 2.36.   CCG

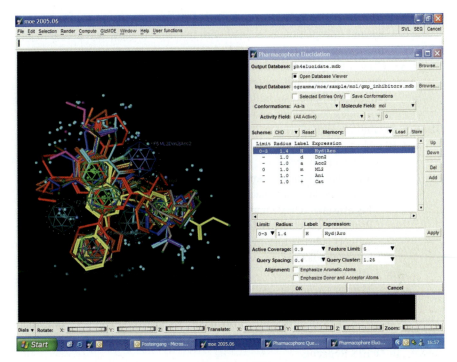

Figure 2.37. CCG

2.9 CHEMINNOVATION SOFTWARE

I. ChemInnovation Software, Inc; http://www.cheminnovation.com/

II. Product Summaries: ChemInnovation Software, Inc provides software for chemistry graphics, molecular modeling, chemical nomenclature, and chemical and biological information management through the Internet. Products include:

  a. CBIS (Chemical and Biological Informatics Systems) provides a suite of Web-based applications for managing chemical and biological materials and data throughout their lifespan. A full CBIS datasheet can be found at http://www.cheminnovation.com/brochures/CBIS.pdf

  b. ChemmInnovation's desktop products are integrated into the enterprise solution CBIS. View more information on CBIS

   i. *Chem 4-D* Draws Chemical Structures Intelligently.

   ii. *Nomenclator Module* assigns systematic names to structures according to IUPAC nomenclature rules.

   iii. *NameExpert Module* understands IUPAC nomenclature rules. If one enters an IUPAC chemical name, it creates the corresponding structure.

    iv. *Chem4D DB Module* manages databases of molecular structures, graphics and information associated with the data. It helps one to search and reuse graphics one has created.

    v. *Chemsite* is a 3D molecular modeling program that allows one to model, animate, render and export 3D molecular graphics for visualization and publication. One can build all types of organic molecules, from small molecules, to proteins and DNAs.

III. Key capabilities and offerings:

  a. CBIS Modules

    i. CBIS Compounds

Manages corporate compounds and their properties. Provides registration procedures for checking duplicate structures and assigning unique Ids. Assigns IUPAC names. Tracks inventory and withdrawal requests. Prints barcode labels for vials or plates. Searching capabilities that support substructure matching or property filtering. Provides links to lab notebooks and assay database.

    ii. CBIS Reactions

Manages reaction schemes and batches. Tracks reactants, products, conditions and reaction progress. Searches reactants and products' structures or properties. Provides links to lab notebooks, compound databases, and reagent inventory.

    iii. CBIS Reagents

Manages reagent inventory. Tracks amounts and withdrawal history. Links to vender databases. Supports multi-level locations and stock room functionalities. Tracks safety information and provide reports for meeting regulations.

    iv. CBIS Biomaterials

Manages databases of proteins, plasmids, phages, other biological materials and their data. Provides links to compounds and assay data. Tools for sequence analysis and searching.

    v. CBIS Sequences

Collections of tools for DNA and Protein sequence analysis, searching and presentation. Supports restriction enzyme mappings, open reading frames, primer design, sequence alignment, property calculations.

    vi. CBIS Bioassays

Manages results for bioassays. Response curve fitting and EC50/IC50 determination. Table pivoting for multiple compounds and targets. Provides links to compounds, notebooks.

    vii. CBIS Bioassay-HTS

Tracks HTS plates and their data. Primary screen tools for identifying active compounds. Secondary screen tools for determining activities.

    viii. CBIS Documents

Manages databases of various types of documents. Supports advanced searching for document contents and abstracts, key words. Automatically tracks document types.

ix. CBIS Reports

Manages project reports or monthly reports. Supports multiple attachments of all types of documents. Provides links to compounds, reactions, and assays. Searches structures, sequences, and properties amount all linked items.

x. CBIS Notebooks

Manages different formats of notebooks. Chem4D can be used to create pages of text, structures, and graphics. Pages are searchable and can be viewed via a browser. Copies of physical notebooks can be stored and viewed on-line.

xi. Chemistry 4-D Draw

Full-featured structure drawing and presentation program. Supports IUPAC nomenclature (names to structures, structures to names). Integrated with CBIS. Supports personal databases of structures, reactions, and graphics.

xii. Safety & Regulation

Maintains compounds and reagents' safety datasheet. Provides reports of safety data for meeting regulations.

xiii. **Software, Hardware Requirements:** Client: PC with Windows NT/2000/XP, Macintosh; Server Hardware: Pentium III or Higher, 256 MB Memory (RAM), 1 GB Disk Space; Server Software: Windows NT/2000/XP Databases: ORACLE, Microsoft SQL Server, Others via ODBC.

b. Desktop Products

i. *Chem 4-D* Graph module that creates multi-line graphs of different styles. It supports non-linear and linear curve fitting, response curve fitting and data analysis. The program allows you to create high-quality structures simply by entering molecular names. It assigns systematic names to structures. It includes a full set of tools for drawing, text and structure editing, and labeling. Many features including over a dozen features related to the drawing program are described on the web.

ii. *Chemsite* is a 3D molecular modeling program that allows one to model, animate, render and export 3D molecular graphics for visualization and publication. One can build all types of organic molecules, from small molecules, to proteins and DNAs. Interactive 3D modeling and real-time animation lets you use molecular building blocks or atom-by-atom construction to visualize complex structures – whether protein, DNA, organic or inorganic – as fully-realized, space-filling entities. One can create and playback movies of molecular dynamics simulations. Analysis features include tools and techniques formerly found only on workstation-level modelers. Full support for stereochemistry, for instance, with dashes and wedges around chiral centers and auto-determination of R and S stereo centers is a part of the ChemSite package. Chemsite provides measurement tools ranging from inter-atomic distance to molecular weight and energy calculation.

1. **Modeling Features** Molecular models may be constructed in a variety of ways. A sketching tool is provided with automatic 2D to 3D conversion for easy building of inorganic and organic molecules. Organic molecules may be built from common organic functional groups and fragments. Peptides and proteins may be built with both D and L amino acids. Single and double stranded DNA/RNA polynucleotides may be constructed with ease.

2. **Visualization Features** Models may be viewed as stick figures, balls and sticks, CPK or with polymers as ribbons or extrusions. With stick figures, molecular models may be rotated in real time with the mouse. The program features ray tracing and texture mapping for the utmost in molecular graphics realism. Rendering styles may be mixed as desired. With high resolution displays, photo realistic display of molecular models are easily created. Images may be printed, saved as .BMP or TARGA files, or exported through the clipboard to word processors and desktop publishing programs. ChemSite brings state of the art workstation quality graphics capabilities to the pc.

3. **Molecular Mechanics Features** ChemSite performs energy minimization and molecular dynamics simulation. Available force fields include Amber, mm2 and the ChemSite's default cm+ force field for accurate calculations with almost any molecule. The program performs real time animation of energy minimization and molecular dynamics simulation with small to medium size molecules. With large molecules such as proteins, movies of molecular dynamics simulations may be recorded to disk and played back for real time animation.

4. **Supported File Formats** ChemSite reads and writes the following file formats:
   (a) ChemSite library (.lib)
   (b) ChemSite system (.syc)
   (c) Brookhaven Protein databank (.ent and.pdb)
   (d) MDL mol file (.mol)
   (e) Mopac Z-matrix (.zmt,.mop)
   (f) Connectivity table (.ct)
   (g) Cartesion A (.cta)

IV. Review: A remarkable range of drawing and visualization tools as alternatives or adjuncts to other drawing packages. IUPAC name-to-structure conversion tools automatically run on large sets of molecules and are available with other packages like ChemDraw and ISIS-DRAW. Logistics such as plate mapping and barcode printing are supported, as well as tools for molecular visualization and simulation. A wider exposure of these tools, as their intrinsic value may be greater than is currently widely known. More obvious integration with smi file formats and other widely used tools would be useful as well. Both product offerings and enterprise solutions are available from Cheminnovation software.

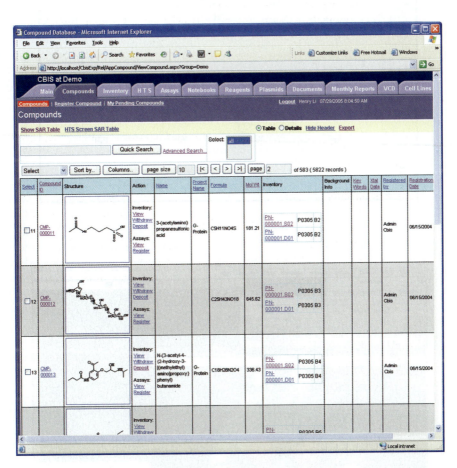

Figure 2.38.   ChemInnovation

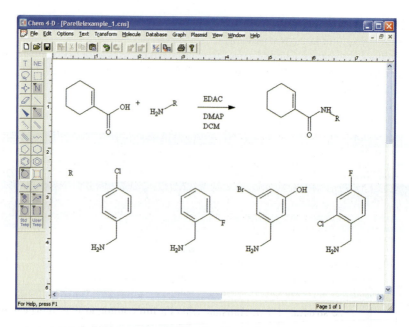

Figure 2.39.   ChemInnovation

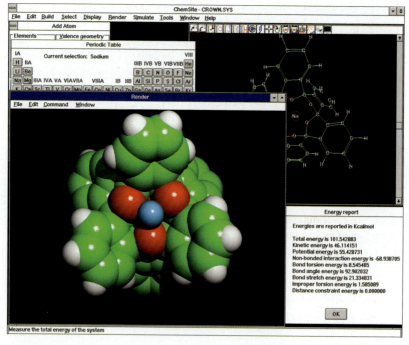

Figure 2.40.   ChemInnovation

2.10 CHEMNAVIGATOR

I. ChemNavigator http://www.chemnavigator.com
II. Product Summaries: ChemNavigator develops and markets scientific software and database technology. The products include 3DPL, CncTranslate, CncMCS, CncDiverse, iResearch Library, On-line iResearch System, and Virtual Screening Services.
III. Key capabilities and offerings:
   a. 3-Dimensional Protein-Ligand Map (3DPL): This service offers a rapid approach for selecting a population of targeted small molecules from starting sets of millions that are likely to bind to a protein surface. 3DPL technology uses a 3D protein structure and large databases of small molecule structures to perform rapid, flexible virtual screening against all likely binding sites on the protein surface. 3DPL uses a patented combination of 3D searching and flexible protein-ligand docking techniques to analyze up to 15 structures per second on a single Windows workstation computer CPU.
   b. CncTranslate: This is a chemoinformatics utility program to perform file conversion among many chemical structure file formats and to calculate properties on chemical structure databases, generate 3D coordinates and normalize structures. CncTranslate offers a full command line interface and may be used as a chemoinformatics utility program within other applications.
   c. CncMCS: This is a maximum common substructure analysis program designed to rapidly identify the maximum common chemical substructure, or common core, contained within a series of chemical structures. The input is a file of chemical structures. The output of the program is a structure file containing the maximum chemical substructure contained within all input structures. The program has many parameters that enable the determination and output of multiple potential common substructures. CncMCS may be used in conjunction with the CncTranslate R-Groups option to determine R-Group variation around a core substructure within a series of related molecules. This is commonly applied in structure activity relationship (SAR) analysis.
   d. CncDiverse: CncDiverse is a software tool for the selection of diverse chemical libraries from within a database of chemical structures. CncDiverse uses fragment based molecular fingerprints to make similarity comparisons between a single molecule and all others in a database. The program uses the Tamimoto Coefficient to compare fingerprints and calculate a similarity index. This index is used discriminate among molecules and select the most diverse (or not similar) set of molecules within the input database. This application is used to select diverse libraries from starting sets of 8+ million structures.
   e. iResearch Library: The iResearch Library is ChemNavigator's up-to-date compilation of commercially accessible screening compounds, building blocks and fine chemicals from international chemistry suppliers. The database currently tracks over 22 million chemical samples. Database licenses include access to regular updates, sourcing information, and ChemNavigator's optional Chemistry Procurement Service. The database

may be licensed on CD/DVD ROM or accessed through an on-line iResearch System subscription.

f. On-line iResearch System: The iResearch System provides access to ChemNavigator's iResearch Library of commercial screening compounds, building blocks, and fine chemicals. Using a single search, once can search all databases for structures of interest, and perform compound selection for specific projects and download result sets for off-line use. The iResearch System also provides detailed information on all chemical suppliers to order products.

g. Virtual Screening Services: ChemNavigator offers a comprehensive set of virtual screening services using proprietary in silico screening technology. Uisng 3DPL ChemNavigator can design a virtual screening program based upon either a 3D model of the target protein or a series of 3D protein targets.

IV. Review:

3DPL, part of ChemNavigator's drug discovery offering, has several advantages over current *in silico* technologies: 600-fold increase in speed of compound docking, entire protein is considered when searching for binding site, on-the-fly flexing of small molecules reduces chance of missing good candidates, and batch processing of large proteomics databases is routine. iResearch Library, part of the chemoinformatics offering, is based on over 23.3 million structures and associated data from 108 chemistry suppliers. Completion of their efforts to screen millions of drug-like compounds against thousands of public and proprietary 3-D protein targets will lead to a new lead discovery database that will provide a continual view of new, potential lead compounds.

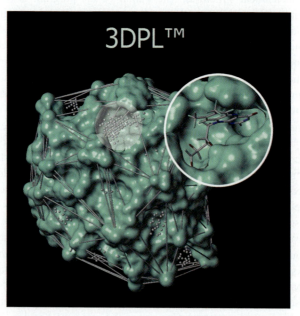

Figure 2.41. ChemNavigator

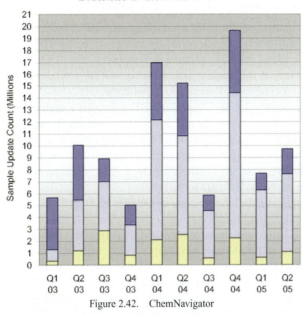

Figure 2.42.   ChemNavigator

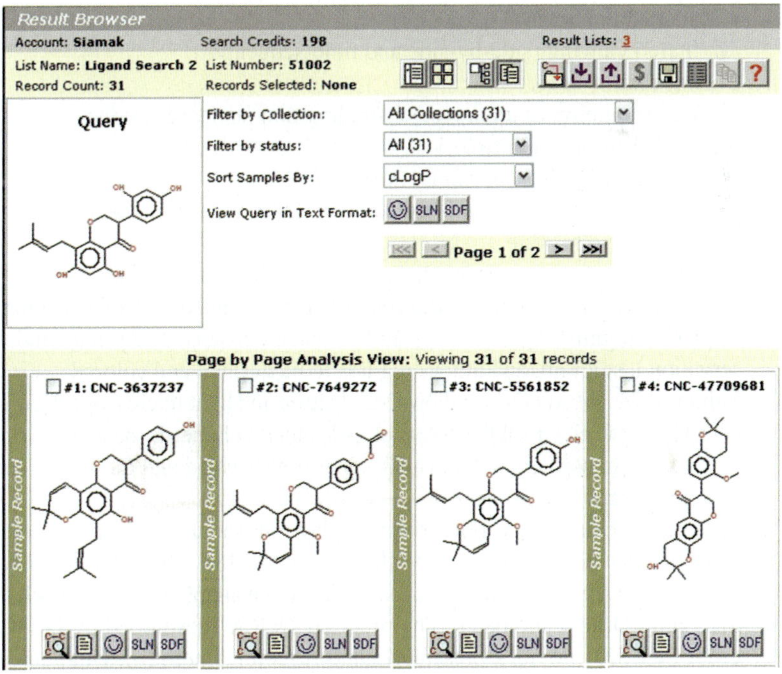

Figure 2.43.   ChemNavigator

2.11   CHIMERA-DOCK-ZINC FROM UCSF

I. Chimera and Dock; http://www.cgl.ucsf.edu/chimera/index.html and the program *DOCK* (developed by the Kuntz group, UCSF) provides possible ligand-receptor binding orientations. http://blaster.docking.org/zinc/ provides a free database of commercially-available compounds for virtual screening. The Chimera extension *ViewDock* allows selecting promising compounds from DOCK calculations and viewing them at the binding site. Also see: UCSF Collaboratory environment: http://www.cgl.ucsf.edu/Research/collaboratory/.

II. Product Summary: UCSF Chimera is a 3-D molecular visualization program available free of charge for academic, government, non-profit, and personal use. For commercial use a license agreement must be obtained. The program can interact with the output of other computational programs and can be customized using Python. Similar terms are available for Dock. ZINC contains over 3.3 million compounds in ready-to-dock, 3D formats.

III. Key Capabilities (for Chimera, see website for latest on Dock and Zinc)
    a. Molecular Graphics:
        i.   Real-time molecular rendering (i.e., wire, stick, ball-and-stick, CPK, ribbons, and molecular surfaces).
        ii. Provides an interface for model translation, scaling, and rotation, and includes a *Side View* viewing tool to adjust clipping planes and scaling.
        iii. Capabilites include interactive *color editing* including transparency.
        iv. Ability to save high-resolution images with high quality for presentations and publications.
        v.   Stereo viewing (side-by-side and time-sequential).
    b. Chemical knowledge:
        i.   Determination of *atom types* including non-standard residues.
        ii. Adds hydrogen atoms.
        iii. High-quality *hydrogen bond identification.*
        iv. Selection of atoms/bonds by element, atom type, functional group, and amino acid residue.
        v.   Interactive bond rotation, distance and angle measurements.
    c. Sequence Viewer: The *Multalign Viewer* extension displays multiple sequence alignments, calculates and shows a consensus sequence and conservation histogram, and allows regions to be defined and colored.
    d. Volume data: *Volume Viewer* shows 3D electron and light microscope data, x-ray density maps, electrostatic potential and other volumetric data. Isosurfaces, meshes and transparent solid display styles are provided and thresholds can be changed interactively.
    e. Molecular Dynamics: Chimera can display molecular dynamics trajectories from several programs (e.g., AMBER, CHARMM, GROMOS, PDB formats). All analysis and display capabilities are available with trajectories.
    f. Screening Drug Candidates: The program *DOCK* (developed by the Kuntz group, UCSF) provides possible ligand-receptor binding orientations. The Chimera extension *ViewDock* allows selecting promising compounds from DOCK calculations and viewing them at the binding site.

g. Collaboratory: The *Collaboratory* is an extension to Chimera that allows researchers share molecular modeling session over a standard network connection. Users have equal control over the molecular structures and all modifications done in a session are instantly seen by all users.

h. Programmability / Extensibility: Chimera is largely implemented in Python with certain components coded in C++. All of Chimera's functionality is accessible through Python and users can implement their own algorithms and extensions without any Chimera code changes allowing their extensions to work across Chimera releases.

IV. Review: UCSF Chimera is an interactive molecular graphics program available for many operating systems (e.g., Microsoft Windows, Linux, Apple Mac OS X, SGI IRIX, and HP Tru64 Unix). Chimera is primarily a molecular visualization tool for small molecules and biological macromolecules. Tutorials and help sheets are both available to assist new users, which is important because the program has many capabilities, and first-time users may not know where to begin; http://www.cgl.ucsf.edu/chimera/docs/UsersGuide/ frametut.html. Chimera's menu-based interface is useful for new and more casual users, while the command-based interface allows experienced users to perform complex tasks quickly. Files of Chimera commands can also be created for later input to the program.

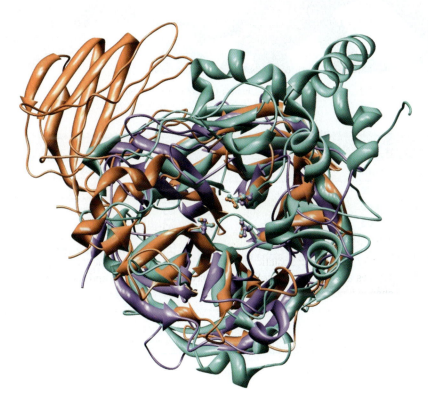

Figure 2.44.   Chimera-Dock-Zinc from UCSF

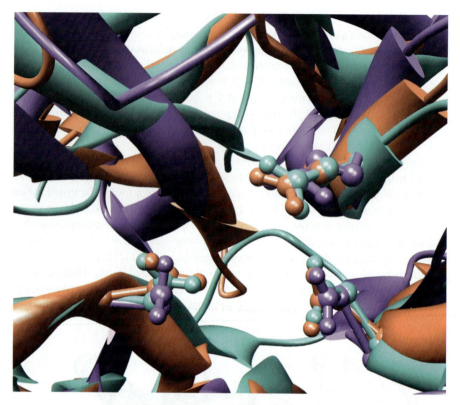

Figure 2.45.   Chimera-Dock-Zinc from UCSF

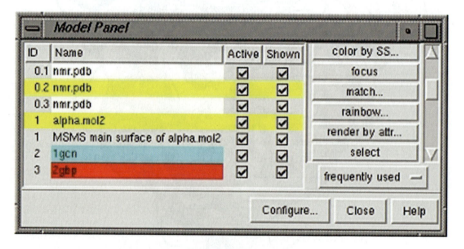

Figure 2.46.   Chimera-Dock-Zinc from UCSF

2.12 COLLABORATIVE DRUG DISCOVERY (CDD, INC.)

I. Collaborative Drug Discovery, Inc.; http://www.collaborativedrug.com/ (for full disclosure: two of the authors were involved in the development of this technology).

II. Product Summaries:

    a. General: Collaborative Drug Discovery, Inc. (CDD) has developed a customizable web-based Molecular Databank for archiving, mining, and collaborating around chemical and biological data. CDD technologies focus on collaborative drug discovery needs especially for groups of researchers to archive and analyze biological data obtained from a range of both low-throughput and high-throughput screens including small molecule enzyme, cell, and animal bioactivity data.

    b. CDD Molecular Databank:

        i. *Archive* experimental data in a searchable, database format

            1. **Web-based interface** to manually enter low-throughput enzyme, cell, PK/ADME/Tox and animal data

            2. **Data "Mappers and Slurpers"** to efficiently batch upload high-throughput enzyme inhibition and high content cell growth assays

            3. **Structured data** typically generated in csv and sd files

                (a) Structured data – such as high throughput enzyme inhibition and cell growth assays typically after analysis in sd files.

            4. **Unstructured data** such as associated original Excel, Word, jpg, gif, tif, and pdf files

                (a) Unstructured data – such as low throughput cell and animal data typically found in excel file and word documents.

            5. **Ability to archive asynchronously** – chemists enter chemical structures and biologists enter biological activity information independently, their results correlate via common structure names.

        ii. *Mine* experimental data – CDD Molecular Databank uses web-based searching incorporating selected components from partner technologies (ChemAxon and OpenEye Software)

            1. **Substructure and keyword searches** including bioactivity, potency and selectivity data, structure-activity relationships, pattern recognition, Boolean searches, and modeling

            2. **Bioactivity potency and selectivity data**

              a. Structure-activity relationships, search, and pattern recognition

              b. Boolean searches (substructure and bioactivity)

            3. **Lipinski's physical-chemical properties and ChemAxon plug-ins** are incorporated for 2D and 3D viewing

            4. **3rd party models and customized algorithms** available via web-based services

        iii. *Collaborate* around data with selected collaborators and colleagues

            1. **Collaborate 3 ways** export or e-mail files or share within the web-based product itself

2. **Collaborate securely** for proprietary data before patents, available to others after patents, publications or non-commercial collaborations
3. **Join and build communities** around therapeutic or target areas
4. **For existing and new collaborations** keep data 100% confidential, share read-only views with others or share globally – with individual control
5. Social networking software for sensitive drug discovery data
6. Each researchers decides if, when, and with whom to collaborate around which data
7. Secure for proprietary data before patents, available to others after patents, publications or for non-commercial, purely scientific collaborations.
8. Acts as an escrow for data prior to sharing
9. Complements open sources of data such as PubChem.

III. Key capabilities and offerings:

a. Archiving: differentiated with the "slurpers" to more efficiently archive both unstructured low throughput data typically in excel files (or other formats) as well has structured (processed) high throughput data typically in sd files. Data can be archived asynchronously so that chemists may enter chemical structures or biologists may enter biological activity independent of each other and correlate results via common structure names. Typically sd files of large molecular files are followed by specific bioassay results for each assay run. Finally, a web-based interface for WYSIWYG (What You See is What You Get) that allow researchers to manually enter or modify any of the archived data.

b. Mining: simple web-based GUI that scientists via the Internet can use to search both chemical (substructure) and biological (potency, selectivity, tox, pk, etc.) data. The technologies are designed for use by a general community of scientists via the web with no training required.

c. (Selective) Collaborating: Unlike most technologies which were designed to be used within a single organization, the Molecular Databank technologies we designed for selective collaborations between organizations. This is differentiated from most existing products which were designed for use behind firewalls. With CDD Molecular Databank technology each researcher can decide if, when, and with whom to (selectively) share their data. Individuals who generate data select who can see and who can modify which data. For those only interested in viewing others data and not sharing any data the technology can be installed behind a firewall to synch up with the web-based ASP version.

IV. Review: Collaborative Drug Discovery Molecular Databank has been designed using primarily open-source systems other than some specific toolkits from OpenEye Software and ChemAxon. Therefore it costs the same to support global communities as it does to support local communities. The technology is designed to support communities of researchers collaborating around specific themes such as common therapeutic areas or targets for academic as well as commercial researchers who need to collaborate around chemical, biological and computationally analyzed drug discovery data.

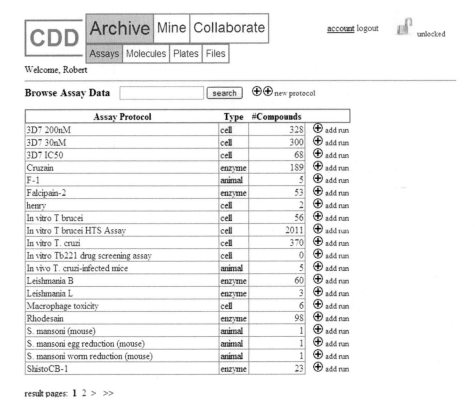

Figure 2.47.   CDD

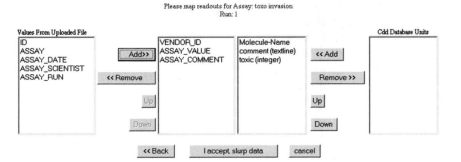

Figure 2.48.   CDD

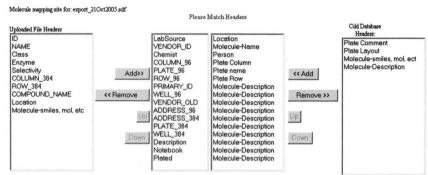

Figure 2.49.   CDD

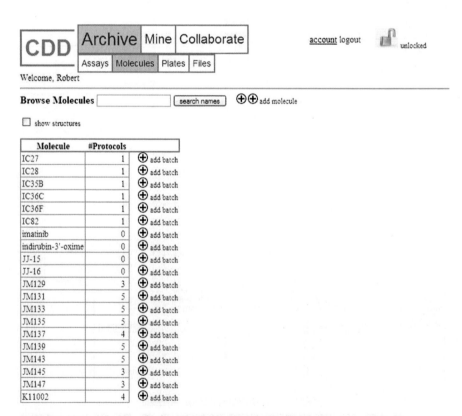

result pages: << < 151.. 161.. 171 172 173 174 175 **176** 177 178 179 180 181 .. 191 .. 201 > >>

Figure 2.50.   CDD

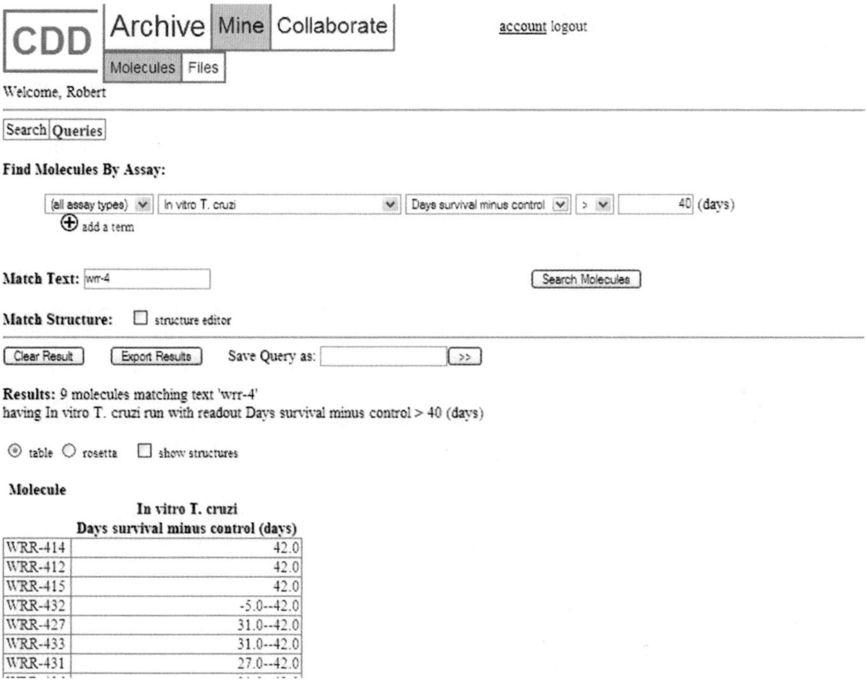

Figure 2.51.   CDD

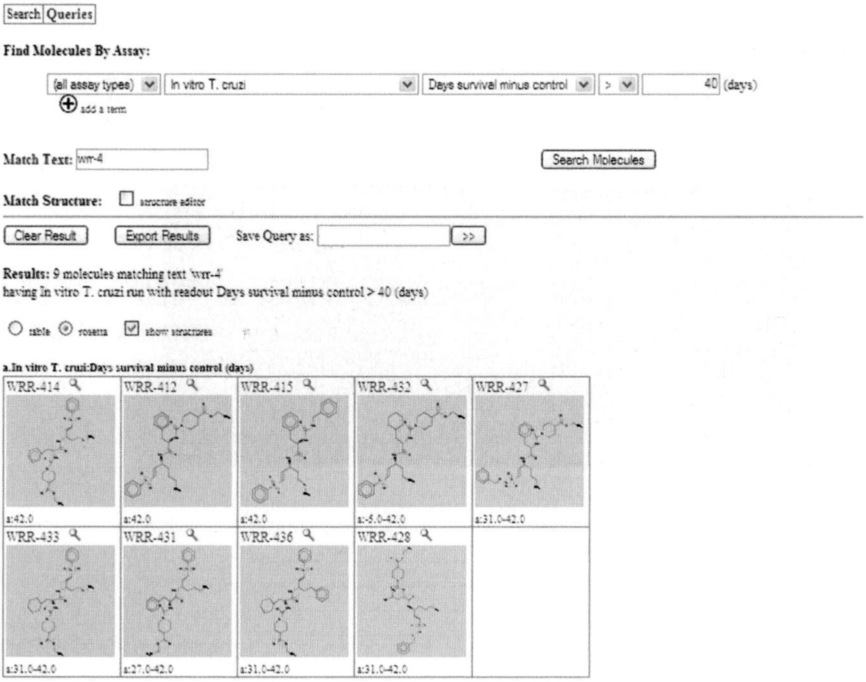

Figure 2.52.   CDD

Welcome, Robert

Edit Hit Colors: In vitro T. cruzi(cell)

≫ **Definition:**

Readouts:

Readout	Type	Description
Comments	textline	Inhibitor concentration, crystals, precipitate, etc.
Days survival minus control	days	Relative days minus control for each run
Days Survival	days	Compound effectiveness is measured by days survival of T. cruzi-infected cells in treated cultures vs. untreated controls

Figure 2.53.   CDD

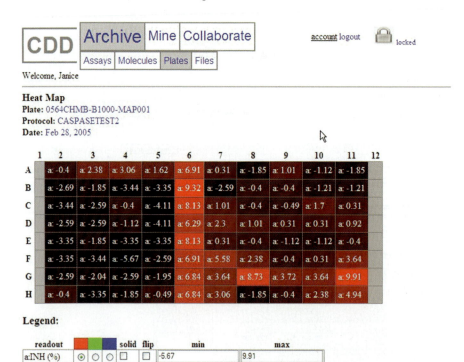

Welcome, Janice

Heat Map
Plate: 0564CHMB-B1000-MAP001
Protocol: CASPASETEST2
Date: Feb 28, 2005

	1	2	3	4	5	6	7	8	9	10	11	12
A		a: -0.4	a: 2.38	a: 3.06	a: 1.62	a: 6.91	a: 0.31	a: -1.85	a: 1.01	a: -1.12	a: -1.85	
B		a: -2.69	a: -1.85	a: -3.44	a: -3.35	a: 9.32	a: -2.59	a: -0.4	a: -0.4	a: -1.21	a: -1.21	
C		a: -3.44	a: -2.59	a: -0.4	a: -4.11	a: 8.13	a: 1.01	a: -0.4	a: -0.49	a: 1.7	a: 0.31	
D		a: -2.59	a: -2.59	a: -1.12	a: -4.11	a: 6.29	a: 2.3	a: 1.01	a: 0.31	a: 0.31	a: 0.92	
E		a: -3.35	a: -1.85	a: -3.35	a: -3.35	a: 8.13	a: 0.31	a: -0.4	a: -1.12	a: -1.12	a: -0.4	
F		a: -3.35	a: -3.44	a: -5.67	a: -2.59	a: 6.91	a: 5.58	a: 2.38	a: -0.4	a: 0.31	a: 3.64	
G		a: -2.59	a: -2.04	a: -2.59	a: -1.95	a: 6.84	a: 3.64	a: 8.73	a: 3.72	a: 3.64	a: 9.91	
H		a: -0.4	a: -3.35	a: -1.85	a: -0.49	a: 6.84	a: 3.06	a: -1.85	a: -0.4	a: 2.38	a: 4.94	

Legend:

readout				solid	flip	min	max
a:INH (%)	⊙	○	○	☐	☐	-5.67	9.91
b:Z-FACTOR	○	○	○				
c:ZPRIME	○	○	○				

Figure 2.54.   CDD

Plate: 0564CHMB-B1000-MAP001
Protocol: CASPASETEST2
Date: 2005-02-28

Scatter Plots

☑ INH (%)

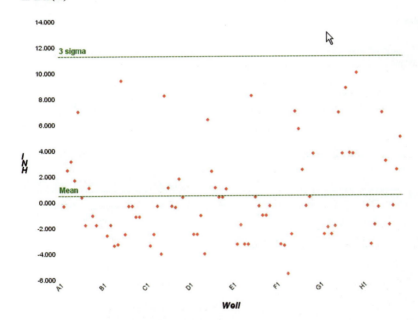

Figure 2.55.   CDD

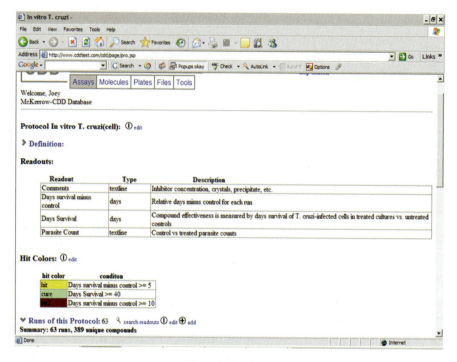

Figure 2.56.   CDD

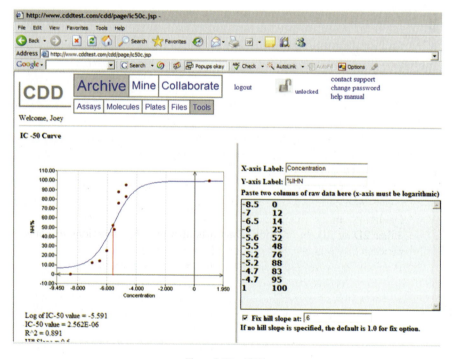

Figure 2.57.   CDD

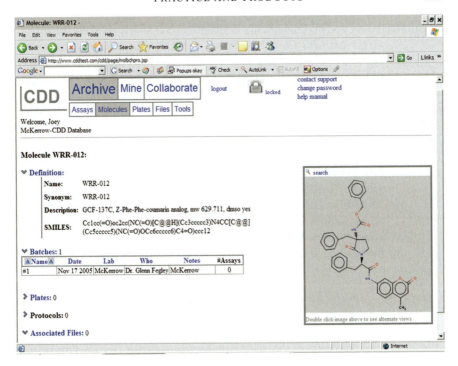

Figure 2.58.   CDD

2.13   DAYLIGHT

I. Daylight Chemical Information Systems, Inc.; http://www.daylight.com/

II. **Products Summary:** Daylight provides chemoinformatics tools and databases for life sciences. Daylight's product portfolio includes DayCart, THOR database, Merlin exploratory data analysis, ACD database, MedChem database, SPRESI database, WOMBAT database, World Drug Index database.

III. **Key capabilities and offerings:**

    a. *DayCart*: DayCart is composed of a set of tools that integrates with Oracle to create, manage and mine molecular and reaction databases. DayCarts uses Daylight's smiles annotation to capture and handle molecular structures in either 2D or 3D. In addition, DayCarts also provides functions to convert between Daylight and MDL formats for import/export.

    b. Database Platform: Daylight offers a proprietary database platform:

        i. *THOR*: THOR is a database system designed to store and retrieve chemical information based on Daylight's SMILES notation. THOR can handle chiral, tautomeric, combinatoric, 3-D, and reaction data, and it can also be customized to work with in-house applications.

        ii. *Merlin*: Merlin is a high-speed exploratory data analysis (EDA) program to mine very rapidly whole THOR databases.

c. Database Content: Daylight has available several widely-used databases:

   i. *MDL's Available Chemicals Directory (ACD)*: The ACD includes several Lipinski's rule-of-5 data (e.g. rotatable bonds, H donors, H acceptors) and it is available in DayCart/Oracle import format and in Daylight's TDT format.

   ii. *MedChem*: The MedChem database contains more than 55,000 compounds, 26,000 activities, 61,000 measured log P values, and 13,900 pKa values among other data.

   iii. *SPRESI*: SPRESI is a chemical literature reference database containing data from 1974 covering almost 3.8 million substances gathered from patents, journals, books, and other sources.

   iv. *WOMBAT*: This database contains over 76,000 chemical biologically active compounds covering over 600 biological targets (e.g., GPRCs, integrins, nuclear hormone receptors).

   v. *World Drug Index:* This database covers about 80,000 biological and pharmacologically active compounds, including all marketed drugs.

d. *Program Extensions*: The modules include:

   i. *Property Package*: This package comprises modules to calculate molecular properties. This package includes:

      1. DayPropTalk: It is an interactive interface that communicates with users or other processes to calculate molecular properties (See DayProp below).

      2. TautomerTalk: It is also an interactive program that produces lists and counts of tautomers.

      3. DayProp: It is a non-interactive program that calculates molecular properties from smile annotations providing the results in Thor Data Tree (TDT) format. DayProp can calculate molecular weight, number of hydrogen bond donors and acceptors, polar surface area, number of rotatable bonds, number of stereocenters, rigidity, and fingerprints *among others*.

      4. Tautomers: It is a non-interactive program that calculates tautomeric structures which may optionally include enol forms.

   ii. *Clustering Package*: This package provides a suite of non-interactive programs to cluster large sets of chemical structures and/or reactions. Parameters to cluster structures include molecular flexibility, similarity, fingerprints, etc. Three types of clustering algorithms are available: JarvisPatrick, k-modes, and sphere-exclusion.

   iii. *PC Models*: This is a software module to access to two chemical models: cLOGP (hydrophobicity partition coefficient) and CMR (molar refractivity).

   iv. *Printing Package*: This is a program set to convert both text or structure data files written by Daylight applications (e.g., DayCart) to PostScript (PS) files.

   v. *Conversion Package*: This package contains a set of programs to convert molecules in MDL formats (e.g., mol files and sd files) to and from Daylight formats (e.g., smiles, TDT).

e. DayCart Partner Solutions: This suite includes:

    i. *ChemCart Web Interface*: ChemCart provides a universal web-based interface to DayCart to search and retrieve chemical structures, reactions, biological data, etc.

    ii. *René Web interface*: René Web interface is another exploratory data analysis interface leveraging and processing data gathered in hitlists.

    iii. *DayBase Windows Interface*: The DayBase Windows Interface is a front end to DayCart and it is capable of importing data from SD, RD or CSV files.

    iv. *Reaction Enumeration System*: This module is a reagent and reaction based enumeration system for parallel and combinatorial planning and synthesis. This module presents a list of compatible reactants for a given reaction with the capability of back tracking starting materials for enumerated products.

    v. *DeltaBook E-Lab Notebook*: DeltaBook is an electronic chemical laboratory notebook to record synthetic reactions, calculate amount of reagents to be used, yields, etc. Searching for data is also allowed.

    vi. *CRISTAL*: CRISTAL is a chemical reagent inventory system for tracking and locating reagents. This system also provides the interface to order chemicals from stockroom or from suppliers.

    vii. *The Modgraph Compound Registration System*: This compound registration system is robust and flexible capable of handling mixtures, stereochemistry, batches and samples, and it can be integrated with other applications.

    viii. *Spotfire DecisionSite*: Spotfire DecisionSite allows the searching and analysis of multi-source, multi-variant data.

f. Developer Tools: For the customization or expansion of modules based on Daylight's technology, Daylight has available a range set of toolkits for developers. The toolkits include SMILES Toolkit, SMARTS Toolkit, Reaction Toolkit, Depict Toolkit, Fingerprint Toolkit, HTTP Toolkit, THOR Toolkit, Merlin Toolkit, and the Program Object Toolkit.

IV. Review:

Daylight provides a comprehensive chemoinformatics product line for life science applications. The THOR database in conjunction with Merlin and ported content databases such as MDL's ACD allows the building, handling, and mining of chemical and biological data using Oracle DB via DayCart and SMILES annotation. The availability of web-based interfaces allows the use and access of Daylight's system from any operating system. And for added flexibility, the availability of toolkits for developers allows not only the customization of an application but also the creation of interfaces or modules as needed. Although not directly made by Daylight, third-party modules allow the chemists to capture and keep track of reaction development in the lab as well as the generation and exploration of virtual compounds for synthesis decisions and planning. The one limitation noted was that third-party modules are not available for the Mac platform.

Figure 2.59.   Daylight

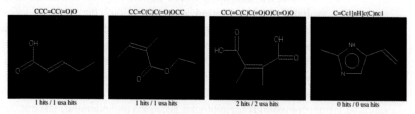

Figure 2.60.   Daylight

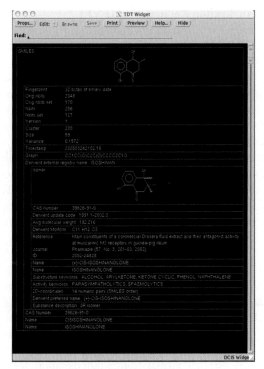

Figure 2.61. Daylight

2.14 EIDOGEN-SERTANTY (PREVIOUSLY LIBRARIA)

I. **Eidogen-Sertanty, Inc. (merger of Sertanty, Inc (previously Libraria, Inc.), and Eidogen, Inc.),** http://www.eidogen-sertanty.com (disclaimer: three of the authors were involved in the founding of Libraria Inc. and development of the ARK/LUCIA technology).

II. **Product Summary:**

   a. **Target Informatics Platform (TIP)**

      i. **TIP** is Eidogen-Sertanty's *structural informatics* platform containing proteome-wide annotation of target structure, binding site, and ligand binding mode information. The TIP knowledgebase contains more than 60,000 high resolution protein structures with annotated small molecule binding sites, including >25,000 non-redundant comparative models of human proteins, covering every major drug target family including proteases, kinases, phosphatases, phosphodiesterases, nuclear receptors, and GPCRs. In addition to TIP's structural content, the knowledgebase also contains pre-computed similarity relationships between sequence, structure, and binding site in the database, enabling large-scale structural and comparative proteomics workflows.

b. **Eidogen Visualization Environment (EVE)**
  i. The Eidogen Visualization Environment (EVE) provides the interface for visualization and comparative analysis of data exported from TIP. EVE enables visualization of Sequence, Structure, Site, and Site-Ligand Contact similarity relationships. EVE also includes several integrated chemogenomic analysis tools, such as **SLiC** for analysis and comparison of Site-Ligand Contacts and **LigandCross** for automated creation of novel ligand scaffolds from ligands known to bind to a target family of interest.

c. **Kinase Knowledgebase (KKB)**
  i. **The Kinase Knowledgebase** is Eidogen-Sertanty's database of structure-activity relationships and chemical synthesis data focused on protein kinases. As of September 2005, the KKB contains over **170,000** SAR data points and >**390,000** unique kinase-modulating molecules, which have been annotated from >**2000** journal articles and patents. All data points and molecules are classified by kinase target name and assay type, along with detailed assay protocols, chemical synthesis protocols, and additional in vitro and animal model efficacy data when available. Integrated into the KKB are target-specific **eScreen** QSAR models built from data in the KKB.

d. **Activity & Reaction Knowledgebase (ARK) (formerly LUCIA)**
  i. The **ARK** platform is a web-based platform to capture synthetic information and structure-activity data that enables scientists to simultaneously build computational QSAR models and explore chemical hypotheses through virtual library enumeration followed by QSAR-based prioritization The ARK high-throughput chemistry knowledgebase has over 1,000,000 unique molecular structures and 14,000 generic reactions with full synthetic procedures for parallel synthesis and high-throughput chemistry. Articles covering both solution- and solid phase-chemistry are organized according to the type of transformation or products generated.

e. **Directed library design using ChIP:** The **Chemical Intelligence Platform (ChIP)** is Eidogen-Sertanty's technology for the "*de novo*" design of focused virtual libraries of small molecules. Within ChIP, reaction transformations contain both empirical reactivity and compatibility information that are dynamically assembled into virtual reaction sequences where commercially available materials can be used as input. Products can then be enumerated using scientist-defined criteria (e.g. QSAR score, pharmacophoric similarity to known actives, structural dissimilarity to undesired chemotypes, etc).
  i. Computational docking and activity assessment using TIP/EVE/eScreen
  ii. Delivery of virtual libraries of molecules meeting specified drug-likeness and predicted activity criteria, along with full synthetic pathways

III. **Key Capabilities and Offerings:**
  a. **Target Informatics Platform (TIP)**
    i. The Calculation Engine used to build the TIP Database consists of:
      1. **STRUCTFAST** – automated comparative modeling

2. **StructSorter** – N-by-N structure-structure alignment algorithm
3. **SiteSeeker** – Binding site prediction
4. **SiteSorter** – N-by-N binding-site similarity algorithm
5. **SLiC** – **S**ite-**Li**gand **C**ontact analysis for binding mode similarity assessments

ii. The TIP database automatically and self-consistently updates itself and computes all relevant similarity information when new protein sequences and structures are uploaded into the system.

iii. The TIP database can be accessed either online through the web-based **TIP Portal,** or deployed behind the company firewall. TIP is integrated with client-side software for data visualization and comparative analysis called the **Eidogen Visualization Environment (EVE).**

b. **Kinase Knowledgebase (KKB)**

i. The Kinase Knowledgebase provides an overview of published knowledge and patents around kinase targets of therapeutic importance, enabling a detailed understanding of the knowledge space around the target of interest and the relevant anti-targets. Inhibitors can be grouped based on structural scaffold to aid in the design of patentable chemotypes.

ii. The KKB helps to prioritize compounds for synthesis or purchase based on their predicted properties. The well-curated SAR data is also useful for the development of proprietary QSAR models for virtual screening. Eidogen-Sertanty has developed nearly 50 kinase-specific **eScreen** QSAR modules for therapeutically important kinase targets. The eScreen training sets are grouped based on common bioassay conditions and common putative binding modes.

iii. The KKB can either be accessed online through Eidogen-Sertanty's ARK platform or licensed for integration into in-house databases.

c. **Activity & Reaction Knowledgebase (ARK, formerly LUCIA)**

i. ARK allows the exploration of virtual chemical space and prioritization of ideas prior to synthesis and screening. Virtual molecules can be entered into the database and molecular descriptors (e.g. HBA, HBD, rotatable bonds, TPSA, cLog P, MWT, etc.) are calculated and the predicted potency and selectivity for the selected target are immediately e-mailed to the scientist. The predicted properties can either be calculated using Eidogen-Sertanty's target-specific **eScreen** QSAR models or any other third party predictive software or proprietary models.

ii. **eScreens** are based on industry standard 3-point-pharmacophore fingerprint technology and high quality empirical SAR data. ARK's predictive eScreen modules for potency and selectivity are specific for a target or gene-family, in contrast to eADME and eTox filters.

iii. The ARK architecture is web-based and is built upon accepted industry-standard database platforms and software (Oracle / Daylight / MDL ISIS).

d. **Chemical Intelligence Platform (ChIP):** To systematically explore chemical space that can be accessed synthetically, ChIP includes over 70 reaction "transforms", encoded with detailed reaction compatibility and reactivity information, which are used for enumeration using commercially available building blocks as input at each step. Crucial information in the form of compatibility and reactivity rules is programmed in to each reaction transform, allowing each transform to select its compatible building block input and the reactive site from any offered reaction input (commercially available building blocks or products from previous reaction steps, etc.). The chemical intelligence embedded in each reaction transform includes the following filters:

   i. **Incompatibility filters** (to avoid side reactions) exclude reactive functional groups, e.g. electrophiles, nucleophiles, acids, etc.

   ii. **Introspective filters** define required reaction center environment (reactivity), e.g. nucleophilic amine, activated aromatic chloride, etc.

   iii. **Global "bad-frag" filters** avoid reactive functionality or undesired motifs, e.g. acylators, undesired elements/isotopes, etc.

e. **DirectDesign Discovery Service**

   i. The DirectDesign Discovery Service combines all of Eidogen-Sertanty's knowledge-driven discovery technologies for the directed *de novo* design of lead compounds and target-focused virtual libraries.

   ii. ChIP is the key enabling technology behind the DirectDesign de novo drug design service (see above description).

IV. **Review**:

ARK (previously LUCIA) allows one to design "Actual-Virtual" libraries based upon 14,000 validated reactions from the high-throughput chemistry literature as templates. ARK is very robust for handling a variety of complex Markush reaction transformations (including stereochemistry, salt forms, and unusually difficult cases). ARK provides and integrated system that handles chemical, biological, and computational models in a single system to help scientists work together in interdisciplinary drug discovery teams. The ability to enumerate virtual chemistry space around actual chemistry space enables one to explore feasible chemical space. The integration of chemistry (reactions), biology (SAR), and informatics (predictions) in one application is unique. The single application for public domain content (Chemistry reactions and SAR) and private domain content is also unique. ARK could benefit greatly if reactions were organized by building blocks and searched in a relational manner. Products enumerated could also be organized into matrices for microtiter plate formats or viewed by building blocks to allow one to visualize the relationship between starting materials and products. SAR from the public domain could be more exhaustive and broader. CHIP provides filters which are potentially predictive of reactivity of building blocks used for virtual libraries. TIP provides visualization across multiple proteins and ligands. ARK, CHIP, and TIP allow the evaluation of synthesizable small molecules, structure-activity relationships, and docking technologies. More details are summarized below:

a. **TIP**

    i. STRUCTFAST automated structure prediction algorithm (see: Dec. '04 CASP6 competition) is integrated into the TIP system and used to generate structural models for the human proteome.

    ii. Integrates sequence information, experimental and predicted structural data, experimental and predicted binding site data, and experimental ligand binding mode information into a centralized database infrastructure, with N-by-N similarity relationships *pre-computed* and stored in the database for easy database querying and retrieval.

    iii. Comparative binding site and binding mode analysis capabilities in client-side EVE software.

b. **Kinase Knowledgebase (KKB)**

    i. Focus on standardization and classification of primary assay data from kinase patents and articles enables efficient navigation of biological relevance of reported kinase inhibitors

    ii. Data readily mined through variety of sources (ARK, Daylight, MDL-Isis), with customizable exporting options to enable easy merging into other programs or workflow management tools

    iii. Integrated eScreens allow supplement of biological assay data with predicted activity data and selectivity analysis

    iv. Released quarterly, with growth rate of >150 publications, $>20,000$ unique molecules, and $>15,000$ SAR data points *per quarter*

c. **ARK**

    i. Integrates chemical synthesis information (reactions), biological activity information (SAR), and informatics (predictions) in a centralized infrastructure

    ii. Processes millions of virtual molecules at the early design stage using $>14,000$ validated high-throughput chemistry reactions, and filter using knowledge-based descriptors and predicted activity from QSAR models

d. **ChIP/DirectDesign**

    i. Designed molecules are synthesizable using commercially available starting materials, and are accompanied with the synthetic roadmaps produced from the ChIP simulation of the literature combinatorial reactions.

    ii. The selection function used for ChIP can utilize either structure-based criteria (e.g. docking score and binding pose analysis) or ligand-based criteria (e.g. pharmacophoric similarity, QSAR scores) if target structural information is not available.

    iii. Ranking and evaluation of initial ranked set of 50–100 novel design candidates in <2 weeks for synthesis and screening.

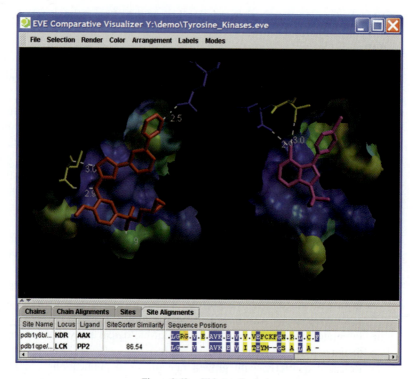

Figure 2.62.   Eidogen-Sertanty

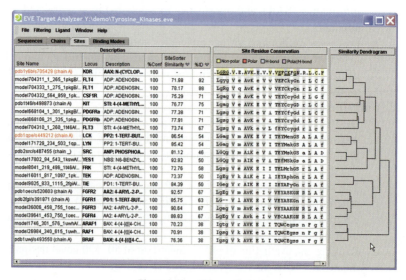

Figure 2.63.   Eidogen-Sertanty

1

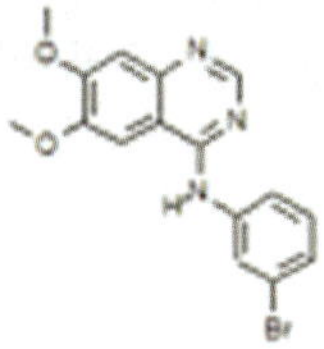	EGFR: Apoptotic Activity (MDA-MB-231)	apoptosis = 31 ±6 % (%Apop at 200uM) apoptosis = 12 ±3 % (%Apop at 100uM) apoptosis = 8 ±4 % (%Apop at 50uM) apoptosis = 10 ±4 % (%Apop at 25uM)
	p56lck Inhibiton	IC50 = 0.06uM
	EGFR Inhibition	IC50 = 0.005nM
	EGFR Kinase Inhibition	IC50 = 0.029nM
	JAK3 Kinase Inhibition (Recombinant JAK3)	IC50 > 300uM
	JAK3 Kinase Inhibition (KL2 Cells)	IC50 > 300uM
	CDK2 Inhibition	IC50 = 250uM
	p38 Kinase Inhibition	IC50 = 6.3uM
	EGFR Kinase Inhibition	IC50 = 0.029nM

Figure 2.64.   Eidogen-Sertanty

Number	Chem ID	Molecule	Measures
1	56	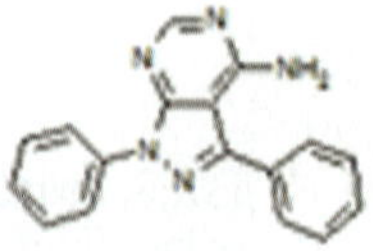	IC50 = 0.80 uM
2	18	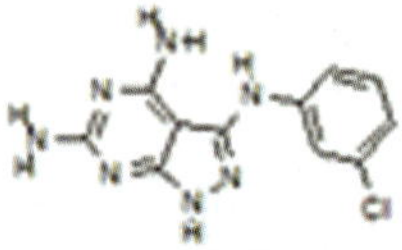	IC50 = 0.22 uM
3	17	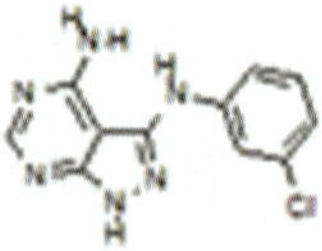	IC50 = 0.16 uM
4	19	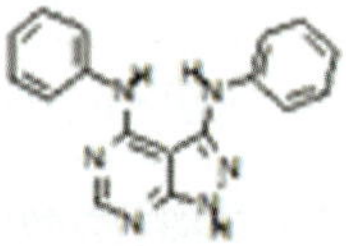	IC50 = 0.058 uM

Figure 2.65.   Eidogen-Sertanty

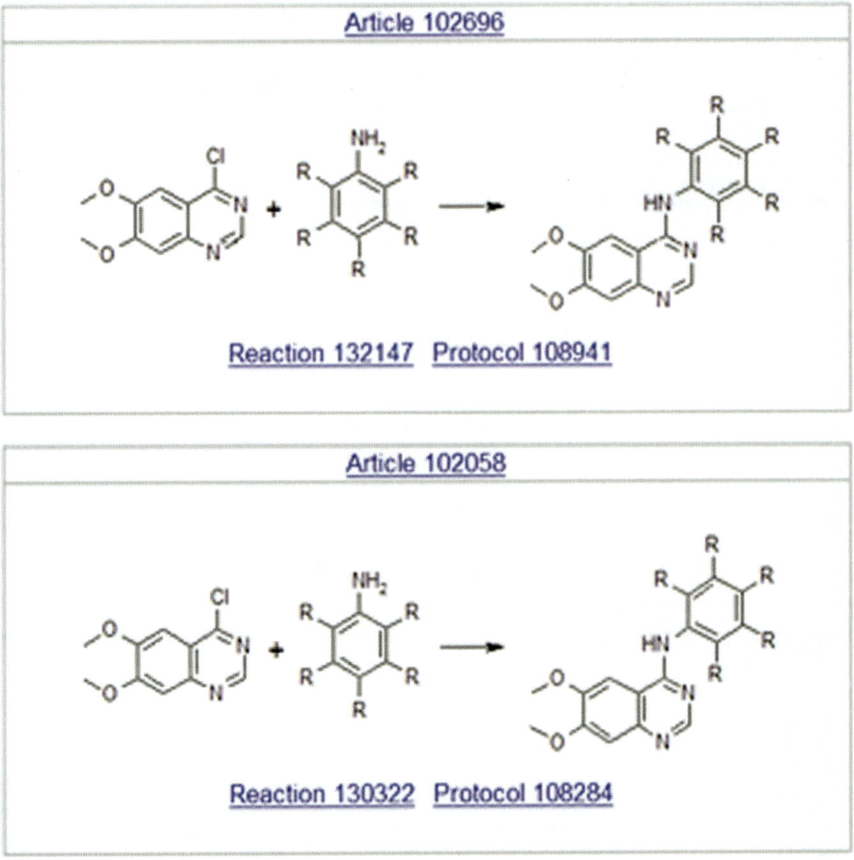

Figure 2.66.   Eidogen-Sertanty

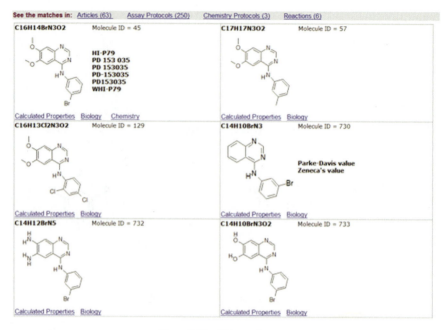

Figure 2.67.   Eidogen-Sertanty

PDB Protein 1xkk (EID 630332)

Add To Project

Title	EGFR KINASE DOMAIN COMPLEXED WITH A QUINAZOLINE INHIBITOR- GW572016
Classification	TRANSFERASE
Compound	MOL_ID: 1; MOLECULE: EPIDERMAL GROWTH FACTOR RECEPTOR; CHAIN: A; FRAGMENT: EGFR KINASE DOMAIN; SYNONYM: RECEPTOR PROTEIN-TYROSINE KINASE ERBB-1; EC: 2.7.1.112; ENGINEERED: YES
Source	MOL_ID: 1; ORGANISM_SCIENTIFIC: HOMO SAPIENS; ORGANISM_COMMON: HUMAN; GENE: EGFR, ERBB1; EXPRESSION_SYSTEM: SPODOPTERA FRUGIPERDA; EXPRESSION_SYSTEM_COMMON: FALL ARMYWORM; EXPRESSION_SYSTEM_VECTOR_TYPE: BACULOVIRUS; EXPRESSION_SYSTEM_VECTOR: PFASTBAC
Date	07-DEC-04
Experimental Method	XRAY DIFFRACTION
Resolution	2.40
Author	E.R.WOOD,A.T.TRUESDALE,O.B.MCDONALD,D.YUAN,A.HASSELL,S.H.DICKERSON,B.ELLIS,C.PENNISI,E.HORNE,K.LACKEY,K.J.ALLIGOOD,D.W.RUSNAK,T.M.GILMER,L.M.SHEWCHUK
Journal Title	A UNIQUE STRUCTURE FOR EPIDERMAL GROWTH FACTOR RECEPTOR BOUND TO GW572016 (LAPATINIB): RELATIONSHIPS AMONG PROTEIN CONFORMATION, INHIBITOR OFF-RATE, AND RECEPTOR ACTIVITY IN TUMOR CELLS.
Journal Reference	CANCER RES. V. 64 6652 2004

Chains					
Chain ID	Chain EID	Sequence EID	# Sites		
1xkkA	630333	629001	4	Add To Project	Protein Search

Sites					
Name	Confidence	#Residues	Chains	Source	Description
s635301	100%	5	1xkkA	PDB Co-crystal	PO4: PHOSPHATE ION
s635302	100%	9	1xkkA	PDB Co-crystal	PO4: PHOSPHATE ION
s635303	100%	28	1xkkA	PDB Co-crystal	FMM: N-[3-CHLORO-4-[(3-FLUOROBENZYL)OXY]PHENYL]-6-[5-([[2-(METHYLSULFONYL)ETHYL]AMINO]METHYL)-2-FURYL]-4-QUINAZOLINAMINE
s635304	89%	55	1xkkA	SiteSeeker	Predicted Site

Figure 2.68   Eidogen-Sertanty

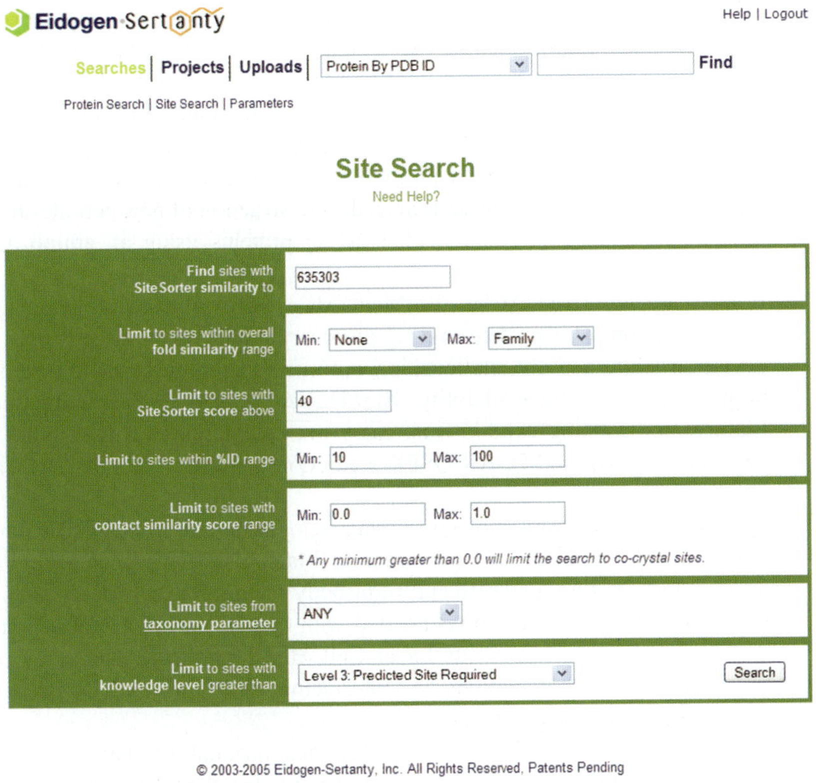
Eidogen-Sertanty
Help | Logout
Searches | Projects | Uploads |
Protein By PDB ID
Find
Protein Search | Site Search | Parameters
Site Search
Need Help?
Find sites with SiteSorter similarity to
635303
Limit to sites within overall fold similarity range
Min: None
Max: Family
Limit to sites with SiteSorter score above
40
Limit to sites within %ID range
Min: 10
Max: 100
Limit to sites with contact similarity score range
Min: 0.0
Max: 1.0
* Any minimum greater than 0.0 will limit the search to co-crystal sites.
Limit to sites from taxonomy parameter
ANY
Limit to sites with knowledge level greater than
Level 3: Predicted Site Required
Search
© 2003-2005 Eidogen-Sertanty, Inc. All Rights Reserved, Patents Pending

Figure 2.69. Eidogen-Sertanty

2.15 Fujitsu Biosciences Group (previously CAChe)

I. Fujitsu Biosciences Group (http://us.fujitsu.com/biosciences)

II. Product Summaries: The BioSciences Group of Fujitsu Computer Systems provides software to help the scientists involved in the design and computational study of compounds with applications in pharmaceutical and agricultural industries. The software programs offered include MaterialsExplorer, MOPAC 2002, and a suite of products under the CAChe name such as Quantum CAChe, BioMedCAChe, and CAChe WorkSystem, etc.

III. Key capabilities and offerings:

a. *CAChe*. (Personal CAChe, Quantum CAChe, Ab initio CAChe, CAChe Worksystem, and CAChe WorkSystem Pro) CAChe is a computer-aided chemistry modeling package for life sciences, materials and chemicals applications with an intuitive property-driven interface that is designed to be used by experimentalists. The CAChe products have a range of computational methods integrated including Molecular Mechanics (MM2 & MM3), Molecular Dynamics (MM2 & MM3), Extended Huckel, ZINDO, DGauss (DFT), and MOPAC 2002. MOPAC handles a broad range of transition metals and molecules of over 20,000 atoms including proteins and polymers. The WorkSystem Pro also comes with the Protein Sequence Editor for the automated alignment of proteins as well as the construction of new peptide structures or homology modeling with existing proteins using an amino acid structure repository.

b. *BioMedCAChe*. (BioCAChe, BioMedCAChe, and ActiveSite automated docking option) This protein-to-drug discovery modeling package is used to devise structure-activity relationships, optimize compound leads to improve both activity and bioavailability. BioMedCAChe can process compound libraries for docking experiments, carry out calculations of descriptors to determine ADME and QSAR/QSPR predictions, and run "rule-of-five" calculations/filters.

c. *MaterialsExplorer*. This molecular dynamics package has been particularly designed for materials science research. MaterialsExplorer can take into account temperature, pressure and rigid-body factors for the calculations. It can import molecular structures created with CAChe and WinMOPAC, predict bilayer interactions via its Layer Cell Modeler module, and carry out high-performance calculations by interfacing with SMP UNIX servers, among many more capabilities.

d. *MOPAC 2002*. MOPAC 2002 is a semiempirical quantum mechanics package operated from the command-line and used to predict chemical properties and reactions in gas, solution or solid-state. MOPAC 2002 incorporates several semi-empirical Hamiltonian methods including MNDO, MINDO/3, AM1, PM3, MNDO-d, and PM5. MOPAC 2002 also includes the linear-scaling MOZYME algorithm for fast calculations of electronic properties of large structures of 20,000 atoms or more such as proteins, polymers, semiconductors, and crystals.

IV. Review:

The suite of CAChe packages can run entirely on Windows or on Macintosh, while the compute engines (e.g MOPAC, ZINDO, DGauss) can also be run remotely on a networked SGI UNIX GroupServer.

CAChe WorkSystem & Pro, and BioMedCAChe provide the necessary tools to execute tasks involved in *in silico* compound screening such as docking, ligand tune-up, etc. MOPAC 2002's MOZYME now allows the execution of semi-empirical calculations of large molecules (proteins, polymers, semiconductors, and crystals) in a quick yet accurate fashion. For the scientists that need to work on molecules whose parameters are either not included in MOPAC 2002 or that need to tune-up particular parameters based on ab initio or experimental data, Fujitsu offers its services to customize parameters on contractual basis.

A stand-alone version of MOPAC 2002 which can be operated from command-line or integrated with other GUIs is available for several other operating systems currently used in industry and academia, including Windows and various flavors of UNIX and LINUX.

It would very helpful all the programs offers were also available for Linux, Unix, Mac, and the new UNIX-based Mac OS. The addition of a de novo ligand builder module would be a nice addition and complement to CAChe WorkSystem & Pro, and BioMedCAChe since the docking capabilities are already implemented in these packages. An engine to create enumerate compound libraries would also fit in their product portfolio and would allow the expansion and better integration of these programs in life science based research programs where combinatorial chemistry and related high-throughput synthesis techniques are currently used.

CAChe Software

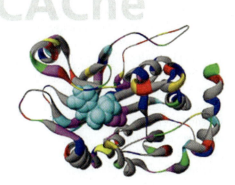

- Built on the solid modeling and property-prediction foundation of CAChe®

- Specialized additional resources for macromolecular analysis, including protein quantum chemistry

- Docking of flexible liganods into proteins with rigid or flexible sidechains

- Correctly models metals, co-factors, solvent, and multiple subunits

- Design paradigms based on QSAR, structural, and pharmacophore-based approaches

Figure 2.70.   Fujitsu Biosciences Group (previously Cache)

BioMedCAChe Software

- *In silico* prediction, modeling, and analysis package designed for the experimental chemist

- Wide range of graphic views and modeling techniques from molecular mechanics through rigorous electronic structure methods

- Model and analyze single molecules or collections, gathering data through property-driven interface in which you designate properties and CAChe® chooses the best calculation method

- CAChe handles all atoms including metals, and chemical groups with all types of binding

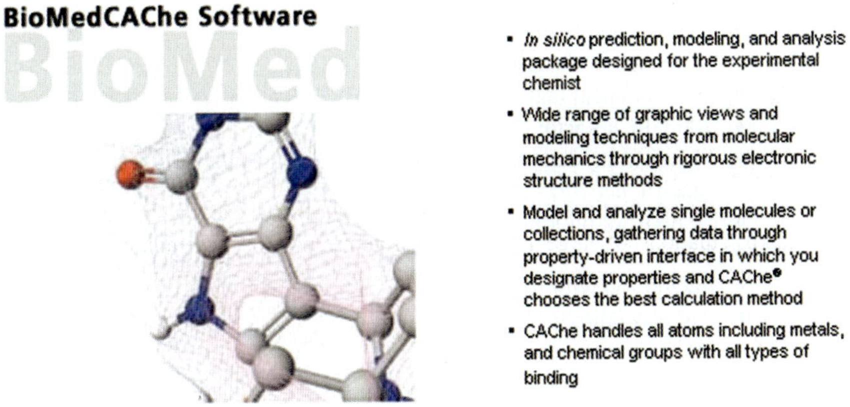

Figure 2.71.   Fujitsu Biosciences Group (previously Cache)

2.16.   GENEGO

 I. GeneGo, http://www.genego.com
 II. Product Summary:

GeneGo's Systems Reconstruction technology connects OMICs and other experimental data to underlying cellular pathways in a way that allows biologists to link
a particular disease condition with individual variations in sequence, gene and
protein expression levels, proteomics content and metabolite composition. This
technology has been implemented into a proprietary computational platform of
content databases, analytical tools and algorithms.

GeneGo has 3 key products: MetaCore for functional analysis of biological
experimental data; MetaDrug for systems pharmacology and MetaBase – direct
access to the content.

 III. Key capabilities and offerings:
   a. Content databases
     i. MetaCore(TM) is an integrative data mining suite consisting of a curated
        knowledge database and an analytical toolkit. The database features a
        comprehensive collection of "small experiments" mammalian protein
        interactions, enzymatic reactions and bioactive chemistry (drugs, metabo-
        lites, toxins). It also contains four hundred unique pathways maps covering
        human signaling, regulation and metabolism. The software enables
        microarray, SAGE, proteomic, metabolomic and assay data to be overlaid
        on the pathways and visualized concurrently. Highlights include:
        • Over 5 million literature and database findings
        • 80,000 direct protein interactions
        • 21,000 human metabolic reactions
        • 2 million synonyms resolved
        • 32,000 pathway/disease links
        • 1,700 transcriptional factors
        • 20,000 drugs and xenobiotics
        • 5,000 endogenous metabolites
        • Over 10,000 human proteins
        • 3,200 human diseases
        • 400 canonical maps
        • 700 GPCR's with ligands
        • Mouse, rat, dog, chicken, chimpanzee, fly, worm, yeast orthologs:
          specific networks and pathways
        • 10 network building algorithms
        • Parsers for expression microarrays, SAGE, proteomics and metabolic data
        • Integration with Resolver and GeneSpring
        • Pathway editing
        • Drugs & drug targets
        • Disease networks

ii. MetaDrug is a systems pharmacology platform used for ADMETox. It has the ability to predict metabolites, prioritize them, then predict affinity for proteins involved in ADME/Tox such as cytochrome P450s, hERG etc. then visualize the predictions in the context of pathways. MetaDrug also contains a pathway database with disease, tissue, localization, ortholog and functional process information, and a suite of software for analyzing microarray, proteomic, metabolomic, SNP and SAGE data that can be overlaid on the pathways and visualized concurrently along with the predicted interactions. To our knowledge no other commercially available tool will allow the user these visualization capabilities for predicted interactions alongside known published interactions Highlights include:

- 70 metabolite rules
- 89 rules for reactive metabolites
- Over 40 QSAR models
- Similarity calculations for each model
- User can add own QSAR models
- Substructure searching
- Generation of molecular properties e.g. Hydrogen bond acceptor and donor counts.
- Export predicted data as sdf, txt and Excel formats
- 15,000 molecules
- 10,000 xenobiotic reactions
- 1,500 substrates with kinetic data
- 2,500 enzyme inhibitors
- Access to the Pathway maps from MetaCore
- Access to the relevant data from MetaCore for ADME/Tox related proteins
- 2 network building algorithms

IV. Review:

GeneGo software allows the visualization of different types of high throughput data within the same system; for example, you can simultaneously relate gene expression, metabolite concentrations, protein levels, protein interactions to molecular entities and pathways. Additionally, you can visualize data on pathway maps and networks. The company also has a GPCR tissue specific database. Signaling cascades and regulatory mechanisms are linked to metabolism, thus increasing knowledge early on in drug discovery phase. The launch of MetaDrug has rounded out the portfolio of products offering a capability for ADME/Tox predictions and analysis of experimental data in a single software suite. This product will be developed with proprietary models for cytochrome P450's alongside an improved metabolite prioritization machine learning algorithm and the addition of rat and mouse data. These technologies are currently being used for the discovery of biomarkers for diseases and toxicity to demonstrate the product's utility. Applications of MetaCore and MetaDrug have been published including their use for modeling nuclear hormone receptor interactions [1], gene-signature networks [2], prediction of metabolites and gene expression [3, 4] as well as analysis of gene expression data from diseased tissues [5]. Further details can be found in these and other reviews [6, 7] listed below:

[1] Ekins S, Kirillov E, Rakhmatulin E, Nikolskaya T (2005) A novel method for visualizing nuclear hormone receptor networks relevant to drug metabolism. Drug Metab Dispos 33:474–481.

[2] Nikolsky Y, Ekins S, Nikolskaya T, Bugrim A (2005) A novel method for generation of signature networks as biomarkers from complex high throughput data. Tox Lett 158:20–29.

[3] Ekins S, Andreyev S, Ryabov A et al (2005) Computational prediction of human drug metabolism. Exp Opin Drug Metab Toxicol in press.

[4] Ekins S, Nikolsky Y, Nikolskaya T (2005) Techniques: Application of systems biology to absorption, distribution, metabolism, excretion, and toxicity. Trends Pharmacol Sci 26:202–209.

[5] Ekins S, Bugrim A, Nikolsky Y, Nikolskaya T (2005) Systems biology: applications in drug discovery. In: Gad SC (ed) Drug discovery handbook. Wiley, New York; pp. 123–183.

[6] Bugrim A, Nikolskaya T, Nikolsky Y (2004) Early prediction of drug metabolism and toxicity: systems biology approach and modeling. Drug Discov Today 9:127–135.

[7] Nikolsky Y, Nikolskaya T, Bugrim A (2005) Biological networks and analysis of experimental data in drug discovery. Drug Discov Today 10:653–662.

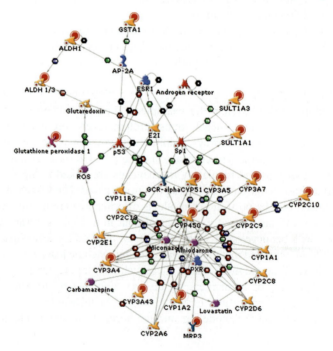

Figure 2.72.   GeneGo

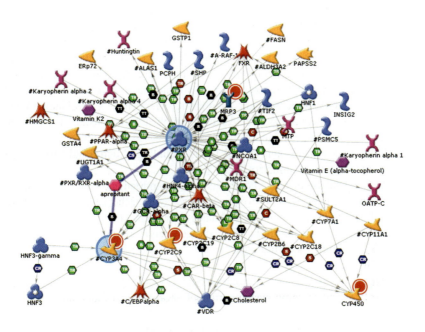

Figure 2.73.   GeneGo

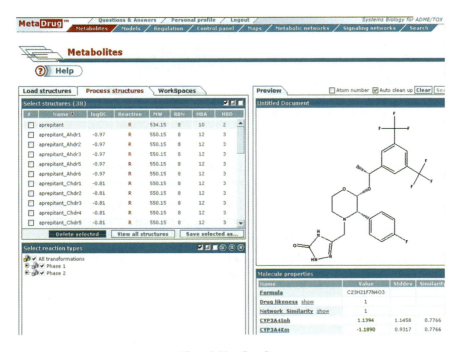

Figure 2.74.   GeneGo

2.17.   GVK-BIO

   I. GVK BIO; http://www.gvkbio.com/
  II. Product Summary
       a. GVK Bio provides various database products with chemical and biological
           data extracted from number of Journal Articles and Patents. These databases
           are available in different searchable formats such as ISIS/Base db-format and
           can be exported into an SD-format, RD-format, ChemFinder, Excel or
           MSAccess as well as Oracle and Web enabled format-using Java as front end.
       b. GVK Bio databases allow users for query based on a number of combinations
           such as numeric and textual fields including structure/sub-structure based
           and molecular similarity based queries.
       c. The database or retrieved results can be exported and resultant 2D structures
           can be converted into 3D structures with other available software for
           Pharmacophore hypothesis, Analogue (3D-QSAR) or Structure based drug
           design (virtual screening) for Lead generation, optimization such others.
 III. Key Capabilities and Offerings:
       a. Services
           1. Custom Curation/Annotation Services
               Chemical, biological, pharmacological and toxicity data
               Protein-Protein interactions, pathways, biomarkers etc
           2. Virtual Screening Models and Focused Libraries
               2D, 3D-QSAR, Pharmacophore based models
               Structure Based models (Docking)
               Consensus Model Generation
               Predictive ADME and Toxicity models
           3. Virtual Screening of Chemical Supplier Compounds
           4. Virtual Screening of Similar/Diverse Compounds
       b. Database Products
           1. MediChem Database consisting over 725,000 records of compounds with
               chemical and biological information on disease, target, and bioassay

	Target	Records	Journal articles	Unique patents	SAR points	Number of Targets
1	Kinase	321,841	1140	3163	552,008	350
2	Protease	177,262	2199	1745	552,000	23
3	GPCR	351,365	5285	3945	863,620	159
4	Ion Channels	123,734	1111	2097	242,377	161
5	Transporters	19,396	562	127	56,743	95
6	NHR	27,266	588	309	47,113	23
7	Phosphatases	9,216	199	65	18,052	57
8	Lipases	8,000	354	171	20,045	21
9	Cytokines	5,000	250	145	12,062	15
10	Phosphodiesterase	4,500	177	47	10,356	19

extracted from 65 Medicinal Chemistry Journals covering a wide range of 5,500 unique targets with 25,00,265 SAR points.

2. Target inhibitor databases with information extracted from both Journals as well as Patents and available for the following protein families as of 2005 with anticipated continuing growth projected:

3. Pharmacokinetic and Toxicity databases (ADME\Tox)

   i. Drug databases (DD) – 1851 records of US FDA approved drugs and Marketed drugs with its Pharmacokinetic parameters, Drug Interactions, CYP interactions, Metabolism, Functional & Binding data.

   ii. Clinical Candidate database (CCD) – 9548 records of compounds which are in the various Phases of Clinical trials with its Pharmacokinetic parameters, Functional & Binding data, Metabolism and Clinical studies.

   iii. Pre-Clinical Candidate database (PCD) – 10,000 records of drug like compounds in the Pre-Clinical stages with its Pharmacokinetic, Functional, Binding data and its Metabolism.

   iv. Mechanism Based Toxicity database – 9000 'drug like compounds' curated from 1035 Journal articles along with its Toxic Mechanism, Adverse effects, Metabolism and Toxicity data.

   v. Toxicity database – 46274 'drug like compounds' curated from 50982 Journal articles with its In vivo & In vitro toxicity along with its Biological data.

4. Target Information database – The Database contains information such as Protein-Protein interactions, Metabolic and Signaling Pathways, Transcription factors, post-Translational modifications, Disease information, various Drugs and Clinical compounds and the Companies of interest.

5. Agricultural Database – The database consists 95493 compounds curated from 7199 unique patents.

IV. Review:

The quantity of quality SAR data and the capability to capture custom data is extremely valuable. The ability to integrate reaction transformations from patent literature would be a useful addition. Generally a combination of medicinal, chemical and computational knowledge is required to have right understanding of scope and limitations of the data necessary for optimum predictive models. Of direct and immediate relevance is knowledge of both the structures and pharmacophores of inhibitors of specific targets. In addition to the McdChcm, Target inhibitor, Agri and Natural Product databases, GVK-Bio also provides Pharmacokinetic parameters databases of FDA approved drugs, Clinical and Pre-Clinical candidates, then Toxicity and Mechanism based Toxicity databases, Target Information database.

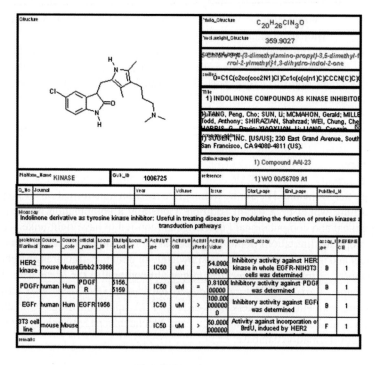

Figure 2.75 — GVK-BIO

	Mole_Structure	$C_{20}H_{23}NO_6$
	MolWeight_Structure	373.4094
	Formula/name	...methyl-4-phenyl-1,4-dihydro-pyridine-2,3,5-tricarboxylic acid 5-isopropyl ester
	SMILES	O=C(C1=C(N(CC)C(=C(C1c1ccccc1)C(=O)O)C(=O)O)OC(C)C
	Title	1) Glucose-Lowering in a db/db Mouse Model by Dihydropyridine Diacid Glycogen Phosphorylase Inhibitors
	Authors	Anthony K. Ogawa; Chris A. Willoughby; Raynald Bergeron; Kenneth P. Ellsworth; Wayne M. Geissler; Robert...
		Department of Basic Chemistry, Merck Research Laboratories, Rahway, NJ 07065, USA; Department of Metabolic Disorders-Diabetes, Merck Research Laboratories...
	claim/sample	1) Compound 1
	reference	1) Bioorg. Med. Chem. Lett., 2003, 13 (20), 3405-3408

Platform_Name: MCD GVK_ID: 449380

S_No	Journal	Year	Volume	Issue	Start_page	End_page	PubMed_ID
1	Bioorg. Med. Chem. Lett.	2003	13	20	3405	3408	14505637

Mode of assay: Dihydropyridine diacid derivative as inhibitor of glycogen phosphorylase - useful in the treatment of type II diabetes

proteinname/Animal	Source_name	Source_code	official_name	Locus_ID	MutType/eLoci	Locus_Pref	ActivityType	ActivityUM	ActivityPrefix	Activity Value	enzyme/cell_assay	assay_type	REFERENCE
HLGPa	human	Hum	PYGL	5836			IC50	nM	=	395.000000000	In vitro inhibitory activity against human liver glycogen phosphorylase a (HLGPa)	B	1
HMGPa	human	Hum	PYGM	5837			IC50	nM	=	1195.0000000000	In vitro inhibitory activity against human muscle glycogen phosphorylase a (HMGPa)	B	1
Rat hepatocyte	rat	Rat					EC50	uM	=	3.500000000000	In vitro inhibition of glucagon-stimulated glycogenolysis in primary rat hepatocytes	F	1

Figure 2.76 — GVK-BIO

	Mole_Structure	$C_{20}H_{26}ClN_3O$
	MolWeight_Structure	359.9027
	Formula/name	5-chloro-...-(3-dimethylamino-propyl)-3,5-dimethyl-1H-pyrrol-2-ylmethyl}-1,3-dihydro-indol-2-one
	SMILES	O=C1C(c2cc(ccc2N1)Cl)Cc1c(c(c(n1)C)CCCN(C)C)(
	Title	1) INDOLINONE COMPOUNDS AS KINASE INHIBITOR
	Authors	TANG, Peng, Cho; SUN, Li; MCMAHON, Gerald; MILLE Todd, Anthony; SHIRAZIAN, Shahrzad; WEI, Chung, Che HARRIS... XIAOYUAN, Li; LIANG, Congxin...
		SUGEN, INC. [US/US]; 230 East Grand Avenue, South San Francisco, CA 94080-4811 (US).
	claim/sample	1) Compound AAI-23
	reference	1) WO 00/56709 A1

Platform_Name: KINASE GVK_ID: 1006725

S_No	Journal	Year	Volume	Issue	Start_page	End_page	PubMed_ID

Mode of assay: Indolinone derivative as tyrosine kinase inhibitor: Useful in treating diseases by modulating the function of protein kinases transduction pathways

proteinname/Animal	Source_name	Source_code	official_name	Locus_ID	MutType/eLoci	Locus_Pref	ActivityType	ActivityUM	ActivityPrefix	Activity Value	enzyme/cell_assay	assay_type	REFERENCE
HER2 kinase	mouse	Mouse	Erbb2	13866			IC50	uM	=	54.0900000000	Inhibitory activity against HER2 kinase in whole EGFR-NIH3T3 cells was determined	B	1
PDGFr	human	Hum	PDGFR	5156, 5159			IC50	uM	=	0.810000000	Inhibitory activity against PDGF was determined	B	1
EGFr	human	Hum	EGFR	1956			IC50	uM	>	100.0000000000	Inhibitory activity against EGF was determined	B	1
3T3 cell line	mouse	Mouse					IC50	uM	>	50.0000000000	Activity against incorporation of BrdU, induced by HER2	F	1

Figure 2.76. GVK-BIO

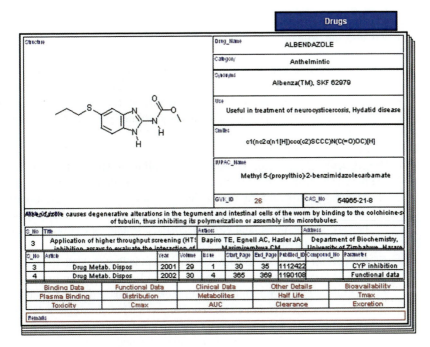

Figure 2.77.   GVK-BIO

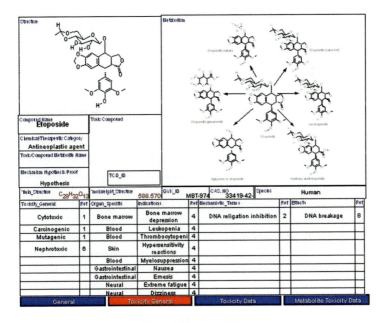

Figure 2.78.   GVK-BIO

2.18   HYPERCUBE

I. HyperCube, Inc., http://www.hyper.com/default.htm
II. Product Summaries:
   Hypercube specializes in molecular modeling software for researchers and
   students. Their principal product is HyperChem, which runs on Microsoft
   Windows and is available in two forms – Standard and Release 7.5. HyperChem
   Lite is a more affordable version and contains a product offering that is tailored
   to students.
III. Key capabilities and offerings:
   a. HyperChem 7.5 has several offerings that allow greater manipulation of
      chemical structures. A brief summary of the new capabilities follows.
      i. Manipulate Protein Structures – dealing with protein structures is easier
         now that HyperChem supports secondary structure descriptions –
         helices, sheets, turns, and coils. Secondary structures can be individual
         selected, colored, and rendered.
      ii. Support for Secondary Structure Information in Protein Data Bank files –
          HyperChem recognizes and supports secondary structure information in
          its molecule files. Information from protein database (PDB) files is cap-
          tured for and retained in HIN files. The peptide builder supports this new
          capability and adds a secondary structure description to all residues.
      iii. Protein Secondary Structure Rendering – Secondary structure rendering
           now includes ribbon lines, narrow ribbon sheets, thick ribbon sheets,
           encompassing helical cylinders and a coil rendering.
      iv. Enhanced Protein Builder Capability – In addition to alpha helices and
          beta sheets, the peptide builder now supports beta turns, parallel and
          anti-parallel beta sheets, left-handed alpha helices, 310-helices, and
          pi-helices.
      v. Large Molecule Electron Density Approximation – A rapid new method
         is available for calculating and displaying the electron density and elec-
         trostatic potential of molecules, thereby allowing users to display the
         electron density of large proteins.
   b. HyperChem 7 Student Edition is a special version of their molecular model-
      ing environment that unites 3D visualization and animation with quantum
      chemical calculations, molecular mechanics, and dynamics. The following is
      a brief summary of the product's features.
      i. Density Functional Theory (DFT) has been added as a basic computa-
         tional engine to complement Molecular Mechanics, Semi-Empirical
         Quantum Mechanics and *Ab Initio* Quantum Mechanics.
      ii. NMR Simulation is available in the form of HyperNMR, a package that
          calculates chemical shifts and coupling constants for molecules as large
          as proteins. Based on a solution of the quantum mechanical coupled-
          Hartree-Fock equations rather than simple database lookup, this package
          allows full exploration of NMR parameters in any situation, such as a

new or novel chemical environment where simple database interpolation is impossible.

   iii. Database Package allows for database search and retrieval of molecules for subsequent molecular modeling calculations as well as the storing of computed properties and optimized structures of your molecules in a new database. Included with the product is a sample database of 10,000 molecules that have previously been optimized with HyperChem.

   iv. Computational Capabilities – HyperChem allows users to explore quantum or classical model potential energy surfaces with single point, geometry optimization, or transition state search calculations.

c. HyperChem Lite 2.0 is an integrated molecular modeling product for researchers, educators, and students. It is a powerful package with capabilities for visualizing, analyzing, and simulating molecules, and for communicating information about molecular structures. A brief description of key product attributes is described below.

  1. Model Building and Visualization

    (a) Rendering choices include all standards formats including: balls and cylinders, sticks, and CPK spheres, with and without shading. Stereo viewing is available, and the orbital and electron densities can be displayed as wire-mesh or shaded solid surfaces, or as contour plots.

    (b) Extensive selection and highlighting capabilities allow users to focus on areas of interest in complex molecules.

    (c) Mouse control of stereochemistry, rotation around bonds, and "rubber banding" of bonds makes manipulation of structures straightforward.

  2. Computational Methods – HyperChem Lite integrates molecular mechanics and semi-empirical quantum mechanics (molecular orbital) calculations into a single package. A brief description of product features and capabilities follows.

    (a) Use the Extended Huckel method to calculate electronic properties and orbitals.

    (b) Use the MM+ force field for general-purpose molecular modeling, including geometry optimization to find stable structures.

    (c) Build in structural restraints to your optimizations.

    (d) Use molecular dynamics playback facilities to analyze chemical reactions and the trajectories of colliding molecules.

    (e) Display orbital energy levels, and the orbitals themselves, as 2D contour plots or as 3D "solid" objects.

d. HyperChem Web Viewer is a free web browser plug-in that allows dynamic 3D views of molecules. Features include real-time rotation and manipulation, animation of normal modes, picking of spectral peaks and more. The Web Viewer can be used to view the results from *HyperChem Professional Release 7* and *HyperChem Professional Release 6*, the *Hypercube Compute Site*, or any of the molecules in the Hypercube online gallery. The Web Viewer provides an interactive 3D visual experience as opposed to the usual static picture.

1. Pocket HyperChem 1.1 is the first chemistry software to run on Windows CE based devices. This product provides the basic molecular modeling and computational chemistry functionality of HyperChem on a portable Palmtop PC platform, allowing the user to work in environments beyond those possible with desktop PC's or notebook computers.

IV. Review:

The HyperChem product line is a powerful software package for researchers, academicians, and students. An appropriate product offering is available to each group. For example, the student version of HyperNMR calculates chemical shifts and coupling constants for complex molecules. The Web Viewer browser allows dynamic 3D viewing of molecules. The Pocket HyperChem 1.1 allows chemists to work on ideas away from the lab or the office through use of a Palmtop PC platform. The Pocket HyperChem 1.1 runs on Windows to provide basic molecular modeling and computational chemistry functionality on a portable Palmtop PC platform. Students are able to perform tasks ranging from building 3D structures from sketches to performing optimizations using semi-empirical methods.

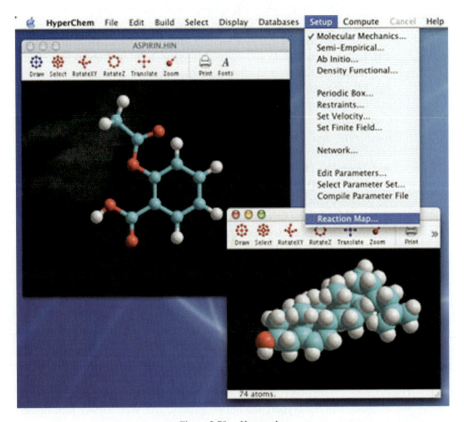

Figure 2.79. Hypercube

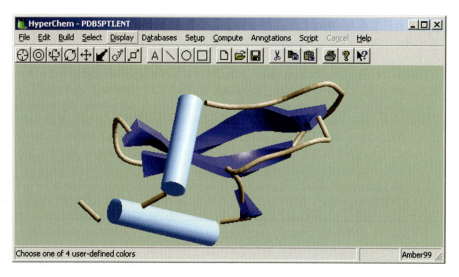

Figure 2.80.   Hypercube

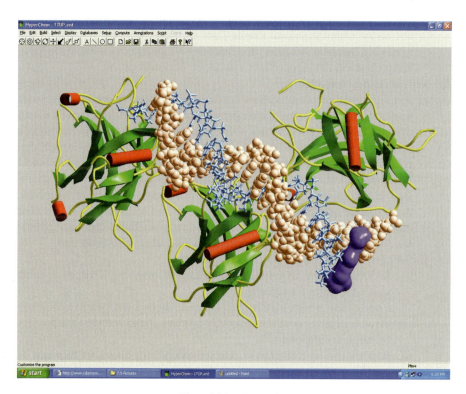

Figure 2.81.   Hypercube

2.19   IDBS

 I. IDBS; http://www.idbs.com/
 II. Product Summaries:
   a. ActivityBase is an integrated chemical and biological data management plat-
      form which includes the AssayBase module for capturing, managing and
      using biological data and ActivityBase for chemical compound registration.
   b. SARgen provides an interface for creating queries of biological and chemical
      data within ActivityBase, and is used for studying structure-activity relation-
      ships and producing SAR reports, as well as compound activity profiles, proj-
      ect summaries and other types of reports.
   c. The Discovery Warehouse consolidates and integrates research data from mul-
      tiple ActivityBase databases, as well as other databases across the enterprise.
   d. ActivityBase Inventory Manager module which provides functionality for
      tracking vials and plates, sample management, request management, request
      tracking and fullfillment management.
   e. IDBS PredictionBase is a suite of applications for building QSAR models.
   f. *XLfit is* a curve-fitting add-on for Microsoft Excel that allows users to per-
      form curve-fitting operations without leaving the spreadsheet environment.
   g. IDBS Chemistry Tools for Oracle, ChemIQ and ChemXtra, provide the tech-
      nology foundation to develop proprietary, highly customized systems for
      chemistry. ChemXtra is an Oracle Data cartridge available for storing and
      retrieving chemical structures and reactions. ChemIQ is a chemical toolkit to
      bring chemical intelligence into a development environment.
 III. Key capabilities and offerings:
   a. ActivityBase Suite: ActivityBase is a suite of advanced discovery experiment
      management applications that enable scientists to capture, manage and use all
      of the data fueling the drug discovery process, from structure registration, to
      screening, to dose response, to behavioral studies. A single, integrated dis-
      covery informatics framework brings together biological data and chemical
      information. Data from different therapeutic and project areas can be organ-
      ized in a consistent, workspace to identify structure activity relationships.
      ActivityBase integrates with Microsoft Excel and Microsoft Word, familiar
      applications for most researchers, and the Oracle relational database applica-
      tion. The software also provides a configurable, defensible audit trail.
   b. SARgen suite within ActivityBase:
      i. SARview assists in formatting and further analyzing SARgen datasets
         into SAR tables for reports.
      ii. ProfileView brings together information about a compound in a single
          view to generate compound activity profiles and to compare the profiles
          of multiple compounds.
      iii. Reports can be restricted to selected samples or plates, or they can be
           based on criteria such as maximum/minimum values or test dates, in any
           combination. SARgen also provides options for displaying results

including: mean values, individual results and related information such as plate and well or experiment IDs.

c. The ActivityBase Inventory Manager (AIM) facilitates the storage and tracking of compounds and containers. It comprises a web-based inventory request application (AIM-web) and a client-based inventory management desktop (AIM-client). Since ActivityBase and AIM share the same data model, inventory and screening workflows.

d. PredictionBase consists of the following applications:

   i. QSAR KBase: KnowledgeBase model generation – allows computational and medicinal chemists to build knowledge models using existing data.

   ii. PharmaExpert: property prediction – enables synthetic and medicinal chemists to make design decisions about how to modify compounds to improve selectivity and potency or reduce toxicity.

   iii. HTSFilter: virtual screening:

     1. Enables medicinal, synthetic and combinatorial chemists to filter large sets of compounds and identify the most promising against a given drug profile.

     2. Allows HTS Screeners to identify required screening experiments and prioritize their work.

   iv. XLfit 4's chart building and data interpretation tools are accessible from Microsoft Excel worksheets and also compatible with other programs like Word and Powerpoint:

     1. XLfit 4 provides preview and browsing tools to fit curves, perform statistical analysis and view and overlay charts.

     2. XLfit 4 provides context-sensitive menus and curve selection trees for your charts with previews to show how individual elements affect the curve fit.

     3. XLfit 4 Uses models such as PA2, global fitting, automatic outlier rejection and others.

   v. eWorkbook captures corporate knowledge in a searchable, traceable electronic repository that is compliant with 21 CFR part 11. Researchers can search and query experimental data, control access and issue task requests to other team members. The eWorkbook:

     1. Captures and protects corporate intellectual property

     2. Fully compliant with 21 CFR part 11, electronic signatures. QA, QC, SOP

     3. Integrates with existing systems, including ActivityBase

IV. Review:

The IDBS flagship product ActivityBase has been wide adopted for handling bioassay data, particularly high-throughput bioassay data. Biologists often handle data in Microsoft Excel worksheets and find compatibility with excel convenient. Mapping the data into this (or any) database project requires specifying multiple variables to map into the database architecture which is not always trivial. ActivityBase, which was originally used for registering small molecules and their corresponding bioactivities, has recently been extended for registering proteins (both wild-type and engineered) and their corresponding activity profiles too. See Vielmetter J, Tishler J, Ary ML, Cheung P, Bishop R. (2005): Data management solutions for protein therapeutic research and development, Drug Discov Today, 10, 1065–1071.

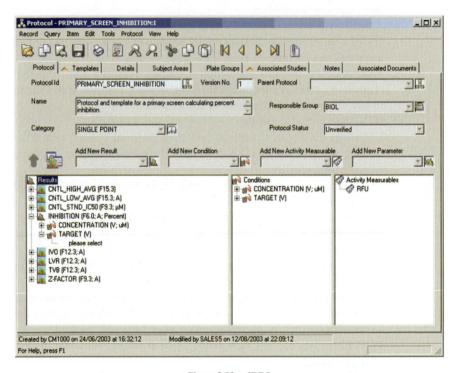

Figure 2.82.   IDBS

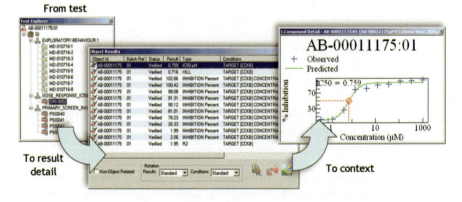

Figure 2.83.   IDBS

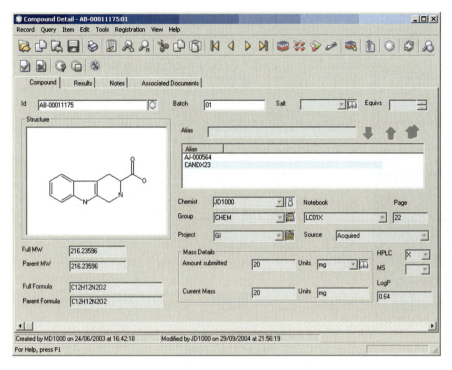

Figure 2.84. IDBS

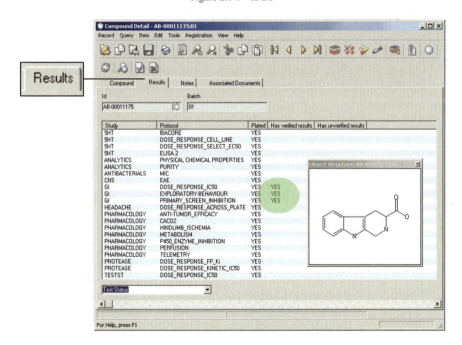

Figure 2.85. IDBS

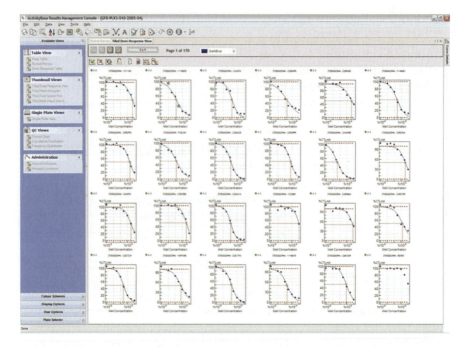

Figure 2.86.   IDBS

2.20   INFOCHEM

I. **InfoChem** (http://www.infochem.de/en/company/index.shtml InfoChem GmbH
    is a software company for chemo-informatics founded in 1989 and based in
    Munich (Germany). InfoChem develops and distributes structure and reaction
    databases as well as software tools required for these applications.

II. Summary:

    The main software tools provided by InfoChem for the handling of databases of
    several million structures/reactions comprise the InfoChem Fast Search Engine
    (IC*FSE*), the InfoChem Chemistry Cartridge for Oracle (IC*CARTRIDGE*), the
    reaction center identification program (IC*MAP*), and the widely used reaction
    classification algorithm "CLASSIFY".

    InfoChem distributes one of the largest structure and reaction database world-
    wide currently containing 5 million organic and organometallic compounds, 3.7
    million reactions and 28 million factual data covering the chemical literature
    published since 1974 (SPRESI). Starting in the 1970s this data collection was
    jointly built by the All-Union Institute of Scientific and Technical Information of
    the Academy of Sciences of the USSR (VINITI) and the German Zentrale
    Informationsverarbeitung Chemie (ZIC), Berlin. The data are continuously
    updated and made available via an Internet or Intranet web-application
    (SPRESI[web]). The data can also be acquired in SD/RD file format.

InfoChem cooperates with several well-known scientific publishing houses in the concept development and implementation of Internet/Intranet versions of printed major chemistry reference works (MRW). InfoChem applications provide standard text search as well as tools to search by structure, substructure and by reactions. Additionally, InfoChem's global Major Reference Work application (gMRW) allows the simultaneous retrieval of structures, reactions and text in several major reference works.

III. Key capabilities and offerings:

    a. **IC*MAP*** is the reaction mapping software tool developed by InfoChem for the automatic determination of atom-by-atom mappings as well as reaction centers in chemical reaction equations. Reaction data may be processed by activating a batch-program as executable and by specifying an RDfile as input. IC*MAP* may also be used as an ISIS/Draw Add-In.

    b. **CLASSIFY** categorizes reactions according to the type of chemical transformation they represent. For this purpose, CLASSIFY first determines the atom-atom mappings and reaction centers (sites) by its own built-in algorithm IC*MAP* and uses this information to define the reaction transformation taking place. Based on this information CLASSIFY assigns to each processed reaction a set of numeric values ("ClassCodes") characterizing the chemical transformation. ClassCodes may be used to cluster large reaction databases, to select subsets of reactions or to define groups of analogous reactions in large sets.

    c. **IC*FSE* (*InfoChem Fast Search Engine*)** is a high performance structure, substructure and reaction search engine that allows retrieval of several million structures and reactions within seconds. IC*FSE* has been developed to handle large chemical databases consisting of millions of structures and reactions like SPRESI or Beilstein (e.g. in DialogLink 5 by Thomson-Dialog). Besides being integrated in SPRESI[web] IC*FSE* is also employed directly in the InfoChem Chemistry Cartridge for Oracle and has been implemented in a variety of chemical environments. IC*FSE* supports the important query features of ISIS/Draw, MDL Draw and ChemDraw.

    d. **IC*CARTRIDGE*** integrates chemical structure and reaction retrieval into relational databases (Oracle) with a high level of functionality. In addition to registration and searching chemical structures can be indexed, edited and normalized. The structures can be handled in standard formats as MOL-File, SDfile and skc-Format. Some of the features of IC*CARTRIDGE* include:

        i. searching for exact structures and substructures in molecule and reaction databases.

        ii. calculation of molecule properties (chemical formula, molecular weight and mass) .

        iii. simultaneous import of structure and associated data from the MDL SD- and RD files .

        iv. normalization and registration of structures.

        v. normalization, auto-mapping and classification of chemical reactions.

    vi. extension of the chemical structure with an additional structure information (addInfo).

    vii. standard database access via extended SQL commands for chemistry searches.

e. **Name Reactions**. InfoChem's tool "Name Reactions" allows the formulation of complex reaction substructure (RSS) queries. The name reactions are organized in hierarchical order based on main reaction categories (addition, substitution, rearrangements, eliminations, oxidations, reductions). Individual reactions can be listed in multiple categories (e.g.: Baeyer-Villiger oxidation in 'Oxidation' and 'Rearrangement') and searched easily by using a tree structure display (point to click) or using a dictionary (text searching). The "Name Reactions" functionality is fully integrated in the SPRESI[web] application.

f. **Synthesis Tree Search (STS)** is a synthesis-planning tool that enables the interactive retrieval of synthesis routes for given target molecules. STS is fully integrated in the SPRESI[web] application and uses the existing 3.7 million reactions reported in the literature to derive the best synthesis path. STS can be used for a given target molecule in two directions:

    i. all published synthesis reactions (preparations) leading to the target.

    ii. all published reactions starting from the target.

    Each molecule obtained in a tree – either as reactant in a synthesis tree or as a product in the forward direction – may be interactively used as a new target and so the tree may be built to any depth and any size required.

g. **SPRESI[web]** is an Internet/Intranet application providing direct access to the SPRESI data: 5.0 million compounds and 3.7 million reactions abstracted from 600,000 references including 165,000 patents. SPRESI[web] has links to several document delivery services (such as FIZ AutoDoc, TIBORDER, CISTI, subito and LitLink for papers and Espacenet, the US Patent & Trademark Office and MicroPatent for patents) and to other chemistry services like ChemNavigator and ACDLabs-Online (I-Lab). Furthermore, SPRESI[web] may be linked to the company proprietary and/or in-house databases by using standard connection table format as crossover key. The user-friendly, highly intuitive web interface allows scientists to easily retrieve molecules, reactions and references. Besides the structure and reaction data SPRESI[web] gives access to over 28 million factual data, containing physical constants, reaction conditions and keywords. In addition to SPRESI[web], the SPRESI data can also be licensed and acquired in computer readable form as SD and RD files.

h. **global Major Reference Works (gMRW)**. InfoChem cooperates with scientific publishing houses such as Thieme Verlag, John Wiley & Sons, Springer, and Elsevier Science in the concept development and implementation of electronic Internet/Intranet versions of printed major reference works (MRW). The software used in these individual electronic versions ("eEROS" from

John Wiley, "Science of Synthesis" from Thieme Verlag, "Comprehensive Asymmetric Catalysis – CAC" from Springer) has been developed by InfoChem and allows the retrieval of structures, reactions and text. InfoChem's application "global Major Reference Works (gMRW)" makes the individual MRWs available in one application that enables global searches over the various MRW databases simultaneously. Currently scientist can perform structure, substructure, and reaction retrieval in approximately 250,000 reactions.

IV. Review:

InfoChem reaction mapping, reaction classification and searching technologies are broadly useful for searching, managing and processing structure and reaction databases. The technology has been used for searching the global Major Reference Works and for integrating content products in MDL's Discovery Gate offerings. InfoChem also offers databases like SPRESI as an alternative to Beilstein. InfoChem's core technologies for handling chemical structures and reactions are used within other products, often without the end user even necessarily knowing.

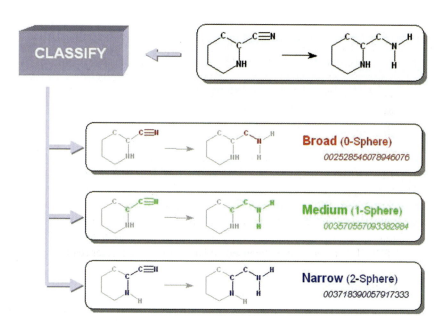

Figure 2.87. Infochem

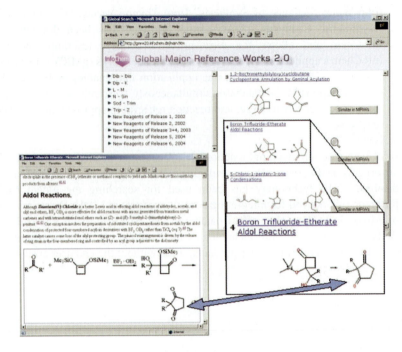

Figure 2.88.   Infochem

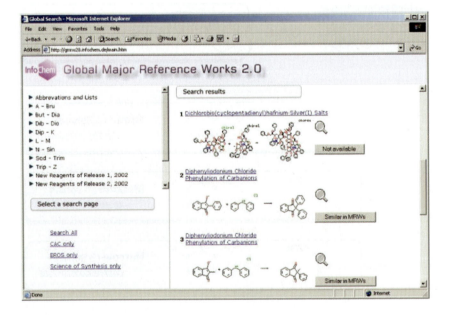

Figure 2.89.   Infochem

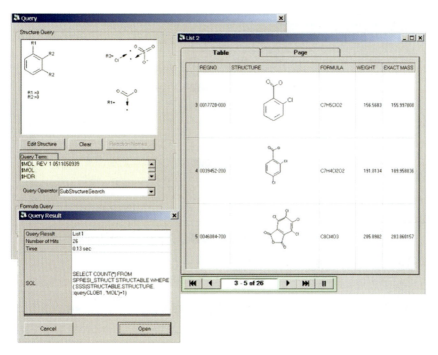

Figure 2.90. Infochem

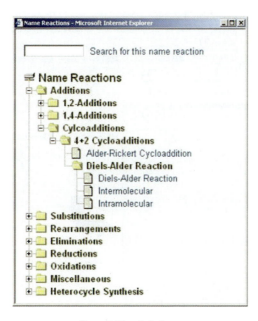

Figure 2.91. Infochem

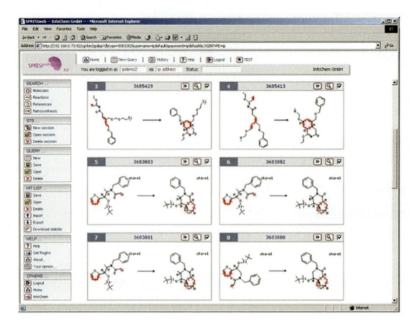

Figure 2.92.   Infochem

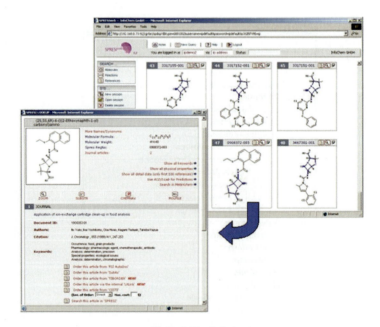

Figure 2.93.   Infochem

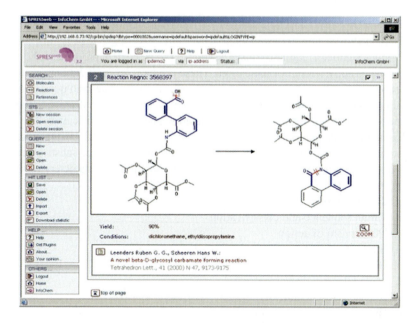

Figure 2.94. Infochem

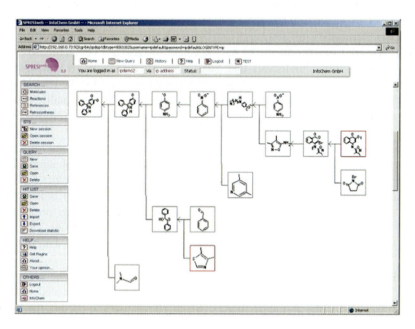

Figure 2.95. Infochem

2.21　JUBILANT BIOSYS

I. Jubilant Biosys; http://www.jubilantbiosys.com/

II. Product Summary: Jubilant Biosys has created some content products in the bioinformatics and chemoinformatics area. These products leverage Jubilant's curation services to incorporate extensive curated databases with structured query modules and front-ends for data retrieval. The content is for the drug discovery process, specifically in the areas of target prioritization and lead identification. The databases are available in Oracle, SD format, and ISIS/Base DB formats and can be exported. The database can be queried across text, structure, substructure and sequences with built-in query modules. Some of the key parameters on which information is curated are:

  a. 2D Structure of the inhibitor

  b. Target sequence and classification

  c. Mechanism of action

  d. Structure activity data

  e. Bioassay

  f. Molecular weight/Formula

  g. IUPAC Nomenclature

  h. Bibliographic information

In addition to Target SAR databases, other databases contain information on ADME and Toxicity parameters, information on proteins and signaling systems involved in the pathways specific genes/proteins, such as Sequence information, SNP details and their functions. Information is derived from public domain data as well as patents and scientific journal articles.

III. Key Offerings:

  a. ***Kinase ChemBioBase***: The Kinase ChemBioBase is a comprehensive database of small molecules that focuses on kinase targets. It contains over 150,000 molecules, and covers over 300 kinase targets for the lead identification process in the kinase area. The database has 175,000 quality-checked SAR (Structure Activity Relationship) points comprising of IC50, EC50, percentage inhibition, ED50, LD50, Ki, Km, etc.

  b. Other Target databases (see: http://www.jubilantbiosys.com/osmd.htm): Jubilant Biosys offers and continues to curate databases of inhibitors of Proteases, GPCRs, Ion channels, and Nuclear Hormone Receptors,

  c. ***PathArt-Pathway Articulator***: PathArt is a pathway articulator, which builds molecular interaction networks from curated databases. This product has information on over 400 regulatory as well as signaling pathways. This product allows users to upload and map microarray expression data onto the pathways.

  d. ***GPCR Annotator***: The GPCR Annotator provides a database that is a repository of information covering areas related to GPCRs. Associated with the database is the annotator module which allows users to predict the family hierarchy of the input sequence as well as function and related information of the sequence.

e. ***Nitrilase and Nitrile Hydratase Knowledgebase***: Nitrilase and Nitrile Hydratase Knowledgebase is a database on these commercially important classes of enzymes to optimize enzymatic reactions and for biotransformation of chemicals.

IV. Review:

Jubliant Biosys provides databases of molecule in a variety of formats. A combination of medicinal chemical and computational knowledge is required to have right understanding of scope and limitations of the data necessary for maximally predictive models. Jubliant's pathway art provides convenient and novel ways for researchers to interact with the scientific literature around biological pathways. In the future, Jubliant plans to integrate the pathway products with the inhibitor products to allow scientists to begin to make hypotheses at the interface of biology and chemistry within complex biological pathways.

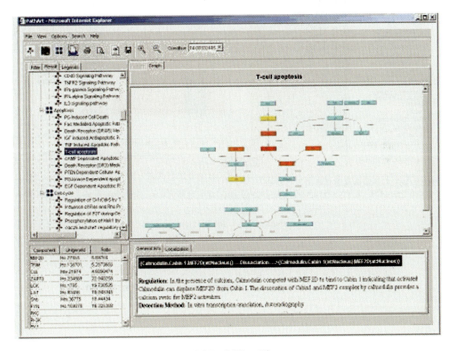

Figure 2.96.   Jubliant Biosys

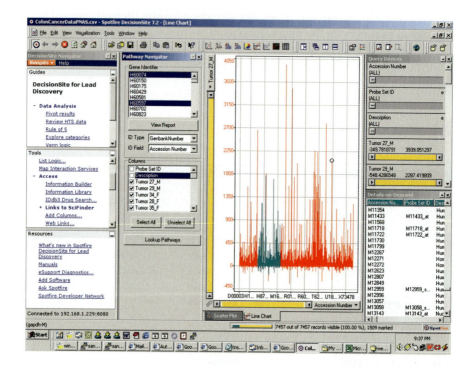

Figure 2.97. Jubliant Biosys

Kinase ChemBioBase

Structure	Mechanism.of.action					ID
	p38alpha kinase inhibitor					6117

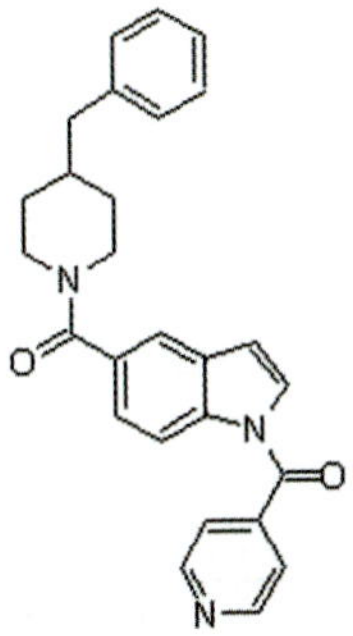

	Assay	Target/Cell_line	Type	Percentage	Concn	Misc_data
1	Inhibition of p38alpha kinase activity studied	p38alpha	IC50	50	0.150µM	-
2	Percentage inhibition of p38alpha kinase	p38alpha	%INH	NA	1µM	-
3	Percentage inhibition of p38alpha kinase	p38alpha	%INH	NA	0.2µM	-
4	Selectivity for p38alpha as	p38beta/p38alpha	Selectivity	-	-	20.1
5	Inhibition of p38gamma, IC50 (µM)	p38gamma	IC50	50	228µM	-
6	Inhibition of Erk2, IC50 (µM)	Erk2	IC50	50	>300µM	-
7	Inhibition of PKA/cAMP	PKA	IC50	50	430µM	-
8	Inhibition of Protein kinase C (PKC), IC50	PKC	IC50	50	>500µM	-
9	Inhibition of cdc2, IC50 (µM)	Cdc2	IC50	50	>500µM	-
10	Inhibition of EGFR, IC50 (uM)	EGFR	IC50	50	>500µM	-
11	Inhibition of p3beta kinase activity studied	p38beta	IC50	50	3.02µM	-
12	Percentage inhibition of p38beta kinase	p38beta	%INH	96	50µM	-
13	Inhibition of DNA-PK, IC50 (µM)	DNA-PK	IC50	50	>500µM	-
14						
15						
16						
17						
18						
19						
20						

Molecular_formula $C_{27}H_{25}N_3O_2$

Molecular_weight 423.5192

IUPAC_Name
(4-Benzyl-piperidin-1-yl)-[1-(pyridine-4-carbonyl)-1H-indol-5-yl]-methanone

Page_No 63 Section Claims

Mol_ID/Patent_Nomenclature
N-(4-Pyridoyl)-(4-benzylpiperidinyl)-indole-5-carboxamide

Article/Patent_Reference
WO 99/61426 A1, Dec 2, 1999

Patent/Article_Title
Heterocyclic compounds and methods to treat cardiac failure and other disorders

Authors_Mavunkel; David Y Liu, George F Schreiner, John
J Dugar, John A Lewicki, John J Perumattam

Institution_Company
Scios; INC; 2450 Bayshore Parkway, Mountain View, CA 94403 (US)

Binding_Site Data not available

Similar_proteins Data not available

Therapeutic_target.3DStructure Data not available

Remarks

JBPL_ID 01012002101 Bio_value Available

QA1 Yes QA2 Yes Curator LVI

Figure 2.98.   Jubliant Biosys

GPCR ChemBioBase ®

Structure	Mechanism of action	ID
	Gonadotropin releasing hormone (GnRH) receptor antagonist	993

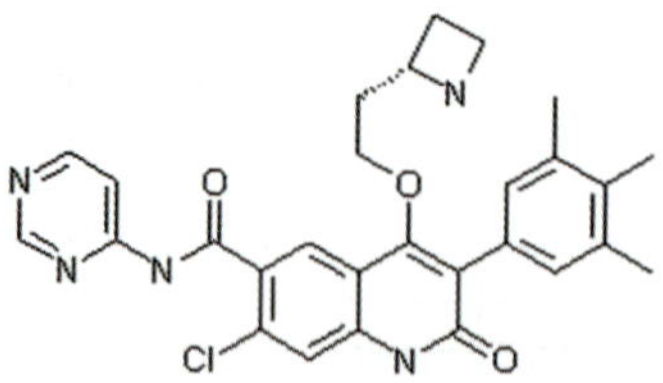

Other targets of action

Therapeutic Indication
Prostate cancer, breast cancer, Endometriosis

Target_Classification
Class A

Reference
J.Med.Chem., 2001, 44(6), 917- 922

Title A potent, nonpeptidyl 1H- quinolone antagonist for the gonadotropin- releasing hormone receptor

Authors: Robert J. DeVita, Thomas F Walsh, Jonathan R Young, Jinlong Jiang, Feroze Ujjainwalla, Richard B Toupence, Mamtha Parikh, Song X Huang, Jason A Fair, Mark T Goulet, Matthew J Wyvratt, Jane- L Lo, **Departments of** Medicinal Chemistry, Biochemistry and Physiology, Pharmacology and Drug Metabolism, Merck Research Laboratories, P.O.Box 2000, Rahway, New Jersey 07065- 0900

Structural_Information   Data not available

*fmla_Structure	*molweight_Structure	Notes
$C_{28}H_{28}ClN_5O_3$	518.0201	

(S)- 4- (2- Azetidin- 2- yl- ethoxy)- 7- chloro- 2- oxo- 3- (3,4,5- trimethyl- phenyl)- 1,2- dihydro- quinoline- 6- carboxylic acid pyrimidin- 4- ylamide

Page_Number	Section	Mol_ID
918	Scheme 1	Compound 1

	Assays	Target	Source	Type	Value		Assays	Target	Source	Type	Value
1	Binding affinity of the ligand at human	GnRH	Human	IC50	0.00044µM	10					
2	In vitro functional antagonist activity of	GnRH	Human	IC50	0.001µM	11					
3	Binding affinity of the ligand at the cus	GnRH	Rhesus monkey	IC50	0.0005µM	12					
4	In vitro functional antagonist activity of	GnRH	Rhesus monkey	IC50	0.007µM	13					
5	Binding affinity of the ligand at rat	GnRH	Rat	IC50	0.004µM	14					
6	In vitro functional antagonist activity of	GnRH	Rat	IC50	0.081µM	15					
7	Binding affinity of the ligand at dog	GnRH	Dog	IC50	0.06µM	16					
8	In vitro functional antagonist activity of	GnRH	Dog	IC50	0.35µM	17					
9						18					

	PK- Assays	Target	Type	Dose	Value		PK- Assays	Target	Type	Dose	Value
19	In vivo area under the curve studied in rhesus	Rhesus monkey	AUC	0.5mg/kg	303ng.h/mL	23	In vivo plasma clearance studied in	Rhesus monkey	Clp	3mg/kg	33mL/min/kg
20	In vivo plasma clearance studied in	Rhesus monkey	Clp	0.5mg/kg	30mL/min/kg	24	In vivo terminal half life time studied in	Rhesus monkey	t1/2	3mg/kg	5h
21	In vivo terminal half life time studied in	Rhesus monkey	t1/2	0.5mg/kg	9.5h	25					
22	In vivo area under the curve studied in rhesus	Rhesus monkey	AUC	3mg/kg	1324ng.h/mL		Bio_Value Available		JBL_ID 08032004090		Curator SRS

Figure 2.99.   Jubliant Biosys

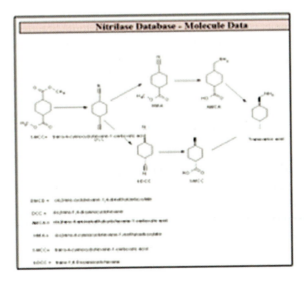

Figure 2.100.   Jubliant Biosys

2.22   LEADSCOPE

I. LeadScope Inc.; http://www.leadscope.com/

II. Product Summaries: LeadScope specializes in comprehensive software solutions to help medicinal chemists deal with the deluge of screening information generated from drug discovery efforts. The product lines primarily focus on data visualization and structural analysis. Content is logically organized according to: structural hierarchy, major structural classes, pharmacophores, protecting groups, spacer groups, and a wide array of functional groups.

III. Key capabilities and offerings:

    a. LeadScope Enterprise allows end users to mine corporate data and visualize the information in a format that is familiar to drug discovery scientists. Enterprise enables scientists in the following areas:

       i. Project Management
      ii. Data Management
     iii. Structure Management
     iv. Hierarchy Management

    b. Personal LeadScope provides tools to assist discovery scientists with the following tasks:

       i. HTS data analysis
      ii. Correlating structural features with biological activity
     iii. Identifying selectivity
     iv. Exploring diversity

      v. Comparing data sets

      vi. Monomer selection for generating combinatorial libraries

    c. ToxScope is a development platform that facilitates the design of safer chemotherapeutics. It has been successfully used to:

      i. Make side-by-side comparisons of efficacy screening and toxocology information

      ii. Use structural alerts from any source

      iii. Explore chemical space of any data set

IV. Review:

LeadScope's data visualization technology makes the interpretation of complex data a routine task. Two-dimensional histograms and scatter plots provide pictorial representations of data sets that can be manipulated using interactive controls. Adjustable sliders, that dynamically reshape data sets, allow users to focus on interesting property ranges. Histograms can also be color-coded to emphasize how compound data correlates to biological activity.

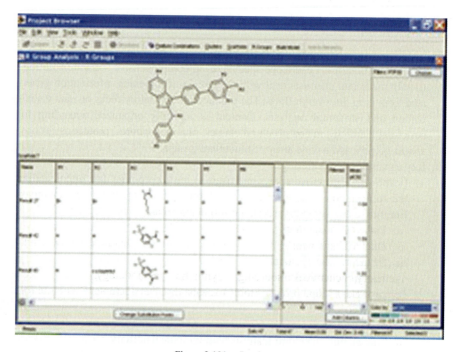

Figure 2.101. Leadscope

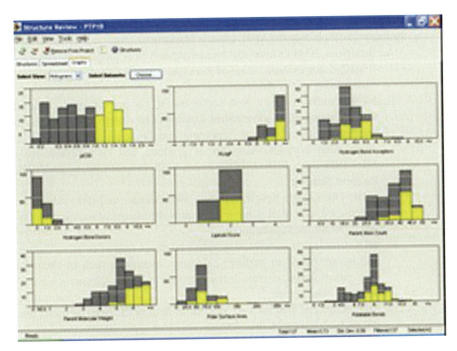

Figure 2.102. Leadscope

2.23   MDL

I. MDL (Elsevier MDL); http://www.mdli.com

II. Product Summaries:

   a. **Framework (including the MDL Isentris platform):** Elsevier MDL's framework offerings include central technology and tools for biopharmaceutical and chemical R&D. This suite includes the new-generation MDL Isentris platform and industry-standard MDL ISIS technology.

   b. **Workflow:** Offerings include a set of applications for managing chemical and biological discovery workflows, including data acquisition, reagent logistics and electronic laboratory notebooks.

   c. **Content and DiscoveryGate:** Content offerings include reference works, literature links, and the world's most comprehensive collection of bioactivity, chemistry, chemical sourcing, synthetic methodology, and EH&S (Environment, Health, and Safety) databases. The DiscoveryGate content platform is an online environment that integrates, indexes, and links scientific information to provide access to compounds and related data, reactions, original journal articles and patents, and authoritative reference works on synthetic methodologies from a single entry point.

III. Key capabilities and offerings:
  a. Framework (Including the MDL Isentris platform):
      i. AutoNom 2000 (Automatic Nomenclature) is a program to generate IUPAC (International Union of Pure and Applied Chemistry) chemical names using drawn chemical structures.
      ii. MDL Central Library is a server-based system for product-based library enumeration and archival software. The system allows access to reagent databases, converts reagents into libraries, and registers the enumerated libraries. Reagents or products can be accessed using MDL ISIS/Base, MDL ISIS for Excel, MDL Project Library, and MDL Reagent Selector.
      iii. MDL Chemscape Server connects Web servers to MDL ISIS/Host. When used with MDL Chime Pro it provides chemical search capabilities in Web-based applications.
      iv. MDL Cheshire is a chemical scripting language for building, validating, and using "rules" to perform particular operations such as chemical convention checks, chemical structure validation, and physico-chemical property calculations.
      v. MDL Chime is free browser plug-in to display 2D and 3D molecules within Web browsers, Java Applets, and Java applications.
      vi. MDL Chime Pro provides Web access to MDL Direct data cartridge technology as well as to a wide range of databases accessible through corporate intranets and the Internet.
      vii. MDL Isentris is the first complete, n-tier informatics architecture for the life sciences, supporting the integration of business processes, data and applications. The new-generation Isentris platform includes four main components:
          1. MDL Base desktop application
          2. MDL Draw chemical drawing and rendering software
          3. MDL Core Interface middle tier integration environment
          4. MDL Direct chemical data cartridge technology
      viii. MDL Base is the desktop component of Isentris that allows scientists to collaborate, share and explore chemical and biological data, while seamlessly moving between applications.
      ix. MDL Draw is a customizable chemical drawing package for drawing and rendering structures and structure queries.
      x. MDL Core Interface provides the MDL Isentris middle-tier data access, user management, object storage and messaging services.
      xi. MDL Direct enables searching and registering molecules and reactions in Oracle. This package includes a chemistry data cartridge for molecule structure management and another cartridge for other for reaction management.
      xii. MDL ISIS (Integrated Scientific Information System) serves as an information management framework for discovery data.

xiii. MDL ISIS/Base is a database management system for storing, searching, and retrieving chemical structures and associated scientific data. ISIS/Base is available as a standalone desktop program on the PC or as a client for MDL ISIS/Host.

xiv. MDL ISIS/Draw is a drawing package to draw chemical structures and sketches. These can be incorporated into documents, Web pages, spreadsheets, and presentations. ISIS/Draw also handles structures in 2D and 3D, polymers, and it comes with templates for common structures such as amino acids, rings, sugars, etc.

xv. MDL ISIS for Excel allows the retrieval and manipulation of biological and chemical data as well as chemical structures in a Microsoft Excel spreadsheet environment. Leveraging the advantages found in Excel, ISIS for Excel permits the creation and mining of databases, construction of SAR tables, exports data directly into databases or SDfiles, and creates reports.

xvi. MDL ISIS/Host provides integrated access to data stored on the server in relational, chemical reaction and 2D and 3D chemical structure databases.

xvii. MDL Report Manager is a reporting tool for biological and chemical data. MDL Report Manager uses MDL ISIS chemical structure searching and live structure reporting to integrate and extract data from multiple databases into pre-formatted reports.

b. **Workflow**

i. MDL Logistics is an application for managing the reagent procurement and inventory process. It combines access to in-house inventory data along with access to chemical supplier catalogs (via MDL Available Chemicals Directory database) and product safety information and MSDS sheets (via MDL OHS database). The system offers regulatory capabilities and support for collections-based procurement for combinatorial chemists.

ii. MDL Notebook is an electronic notebook for capturing, authenticating, integrating, reporting and sharing scientific data. Its scope covers authentication, sign/witness, audit, record repository and reporting functions and supports synthetic chemists with single-, multi-step and parallel syntheses.

iii. MDL Plate Manager is a management system for tracking and sorting data associated with plates and samples. Plate Manager can create plates and samples, reformat plates, import plates and samples, search for plate and sample information, manage plate inventory, and track history.

iv. MDL Assay Explorer is a biological data management system that allows biologists to capture, analyze and store all of their experimental results within a single environment. It extracts data from user-selected files, applies the layout, and calculates the experimental results. Assay Explorer's Dictionary Manager uses a standardized vocabulary for assays, results, materials, and other information. Assay Explorer also uses standard statistical procedures for data analysis to automatically fit large datasets to a wide range of non-linear curves, while acquiring parameter values for kinetic or dose-response studies such as EC50s, IC50s, DT50s, LD50s and NOELs.

v. MDL ChemBio AE is a storage and retrieval system for chemical and biological data of single compounds or entire libraries. ChemBio AE automatically checks for duplicates, strips salts and solvents, assigns corporate ID numbers, and calculates chemical properties.

vi. MDL Reagent Selector is a tool to select reagents from in-house inventories and commercial suppliers. Criteria for selecting compounds can include desired or undesired functionalities, elements, suppliers, and more.

vii. MDL Carcinogenicity Module is used to predict long-term carcinogenicity risk to humans using rodent data. This module is structure based using more than 1300 compounds to build predictive modules and was developed in collaboration with the US FDA.

viii. MDL QSAR is a modeling system to establish quantitative structure-activity and structure-property relationships and create new calculators for *in silico* screening. MDL QSAR uses over 400 built-in 2D and 3D molecular descriptors and a Genetic Algorithm to automatically choose the best set of descriptors to create a model from a given dataset. MDL QSAR comes with data analysis tools to cluster data, visualize in up to 4 dimensions for principal components, identify outliers, and select/deselect clusters for models.

ix. MDL Sculpt is a desktop 3D visualization system that automatically generates low-energy 3D conformation when compounds are entered from MDL ISIS/Draw, MDL ISIS/Base, MDL ISIS for Excel, or MDL Chime. Sculpt provides high-quality visualizations aligning compounds to each other, add solvent-accessible surfaces, protein ribbons, contact surfaces, and CPK and ball-and-stick rendering.

c. **Content and DiscoveryGate**

i. CrossFire Beilstein is a large database for compounds, reactions, properties and citations covering over 9 million compounds and 9.5 million reactions. The data in the database includes citations, titles, and abstracts, which are hyperlinked to the substance and reaction domain entries.

ii. CrossFire Gmelin is a large database of inorganic and organometallic chemistry containing over 1.9 million compounds and 1.5 million reactions.

iii. MDL Comprehensive Medicinal Chemistry is a database of biologically active compounds derived from the Drug Compendium in Pergamon's *Comprehensive Medicinal Chemistry* (CMC). The CMC database includes 3D models and relevant data such as drug class, log P, and pKa values for over 8,400 pharmaceutical compounds (1900-present).

iv. MDL Drug Data Report (MDDR) is a database covering the patent literature, journals, meetings and congresses. This database contains over 132,000 biologically relevant compounds with updates adding about 10,000 a year to the database.

v. NCI biologically active compounds contains over 213,000 structures with corresponding 3D models generated using CORINA 2.4. The NCI database includes a Plated Compounds database, an AIDS database, and Cancer database with inhibitors tested in several of human tumor cell lines.

vi. MDL Available Chemicals Directory (ACD) is an electronic, structure-searchable database of commercially-available chemicals, with suppliers providing information on product purities, forms, grades, available quantities, and prices.

vii. MDL Screening Compounds Directory is a supplier database containing over 2 million commercially available drug-like compounds.

viii. DiscoveryGate is a scientific content platform that provides online access to primary literature sources (including 20,000 journal titles and patent publications), secondary databases (containing over 16 million molecules from Elsevier MDL and partner companies), and integrated major reference works from leading publishers.

ix. MDL CrossFire Commander and MDL® CrossFire Server are applications for searching CrossFire Beilstein, CrossFire Gmelin and the MDL Patent Chemistry Database.

x. OHS Reference provides EH&S data for 18,000 unique chemicals and 165,000 trade names/synonyms.

xi. MDL Metabolite Database comprises a structural metabolic database with entries for particular parent compounds. This database can be used to create, edit, and register metabolic schemes into corporate databases.

xii. MDL Toxicity Database contains toxic properties of over 158,000 chemical substances containing data from *in vivo* and *in vitro* studies of acute toxicity, mutagenicity, skin and eye irritation, tumorigenicity and carcinogenicity, reproductive effects and multiple dose effects.

xiii. ChemInform Reaction Library provides solution-phase synthetic methodology from 1900-present (223,700 literature references) covering new reactions and syntheses including enzymatic and microbial processes.

xiv. Derwent Journal of Synthetic Methods is a compilation of chemical reaction literature from international journals and patent sources covering literature from 1980 to present.

xv. MDL Reference Library of Synthetic Methodologies covers novel functional group transformations, metal-mediated chemistry, and asymmetric syntheses.

xvi. MDL Solid-Phase Organic Reactions (SPORE) covers Solid-phase synthetic methodology from 1963-present. This database is updated quarterly covering solid-support synthesis, development and use of linkers and spacers, functional group protection/deprotection, all the data with literature references.

xvii. ORGSYN database contains an electronic version of the entire series of *Organic Syntheses* (first published in 1921) covering general synthetic methods, product purity, product yield, and hazards, as well as references to the original procedures and journal sources.

xviii. MDL Patent Chemistry Database indexes chemical reactions, substances and related information from World and European patent publications since 1978 and U.S. publications since 1976.

xix. xPharm is an online pharmacological database that links together information on agents (compounds), targets, disorders and principles.

IV. Review:

Elsevier MDL is one of the industry leaders in databases and compound registration systems for the pharmaceutical industry. Because Elsevier MDL has so many products and customers, the products tend to integrate better with MDL's products than with third party vendors. Their new products are built on the open Isentris platform, which is designed to integrate third party products and accommodate existing applications. Third party databases and content sources are integrated through the DiscoveryGate platform.

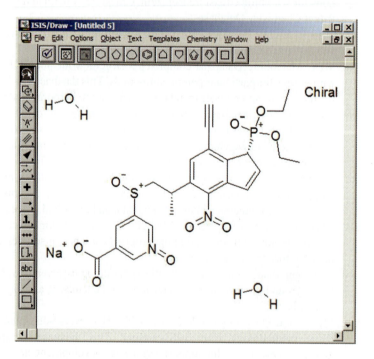

Figure 2.103. MDL

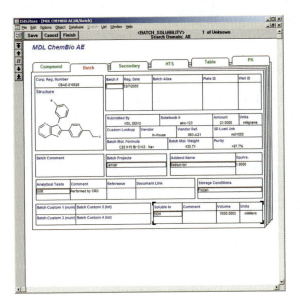

Figure 2.104.   MDL

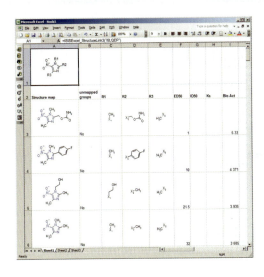

Figure 2.105.   MDL

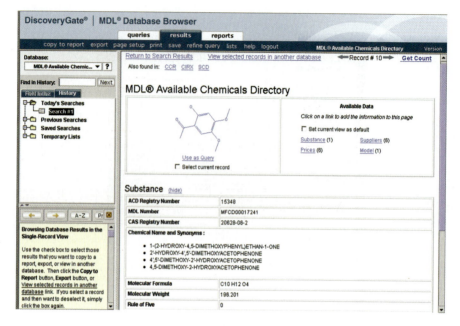

Figure 2.106.   MDL

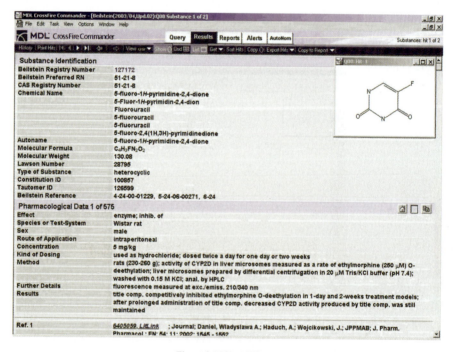

Figure 2.107.   MDL

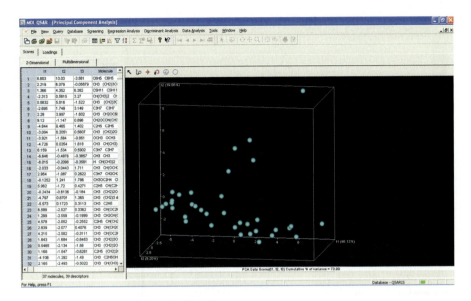

Figure 2.108.   MDL

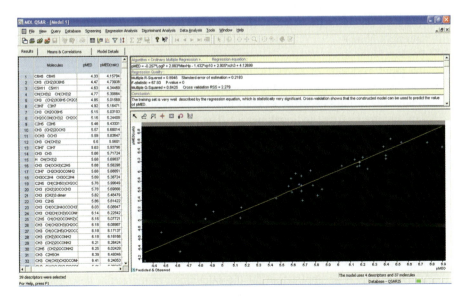

Figure 2.109.   MDL

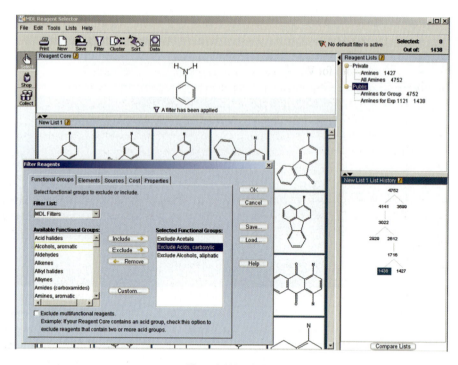

Figure 2.110.   MDL

2.24   MILANO CHEMOMETRICS AND QSAR
RESEARCH GROUP

I.   Milano Chemometrics and QSAR Research Group (http://www.disat.
unimib.it/chm/)

II.   Product Summaries: Milano Chemometrics aims to increase the opportunities for
the rapid exchange of scientific information on molecular descriptors, QSAR
approaches, chemometric methods, and recent developments and results in these
fields. The group's expertise extends beyond this basis to include reference data
sets, environmetrics, and qualimetrics. The Milano Chemometrics site also has
numerous useful links to on-line software where researchers can run common
QSAR algorithms and methods to analyze their own data. The following is a sampling of the sites accessible through Milano's homepage: neural networks and
algorithms for calculating lipophilicity (log P), prediction of molecular properties
for Log P and Log W, and many different biological activities. Their main product
offering consists of Dragon, MobyDigs, Koala and Dolphin. Moreover, Milano
Chemometrics gives the scientific support to Dragon and MobyDigs, which are
commercialized by Talete srl (http://www.talete.mi.it)

III.   Key capabilities and offerings:

a. DRAGON is an application for the calculation of molecular descriptors – a
number extracted by a well-defined algorithm from a representation of a

complex molecular system. These descriptors can be used to evaluate molecular structure-activity or structure-property relationships, as well as for similarity analysis and high throughput screening data analysis. DRAGON is available both for Windows and Linux operating systems.

b. The MOBYDIGS software package has been developed for the calculation of regression models by using genetic algorithms for variable selection to obtain an optimal subset of predictive models. The software is IBM PC compatible and basic operating systems requirements apply (Microsoft Windows 95/98/ME/2000/XP, WINDOWS NT 4.0 or above).

c. KOALA uses Kohonen maps and counter-propagation principles that have been extensively used in neural network simulator software.

d. DOLPHIN deals primarily with optimal design-based Kennard/Stone and Todeschini/Marengo algorithms, which have been successfully utilized in the analysis of both Kohonen and counter propagation artificial neural networks.

IV. Review:

Dragon is very useful for converting data sets into descriptors for complex molecular systems. This ultimately allows one to develop a greater understanding of complex molecules by breaking the overall system into components that can be compared to known systems. The remaining software programs expand researcher's abilities to include applications in artificial neural networks. It would be helpful if the authors allowed downloadable software demos for KOALA and DOLPHIN to enable researchers assess the computing capabilities of these programs.

Figure 2.111.   Milano

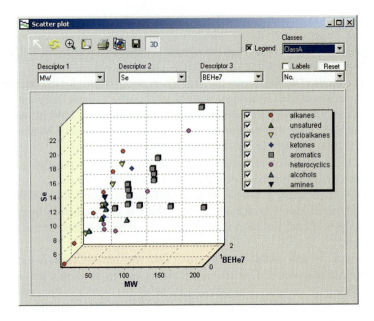

Figure 2.112. Milano

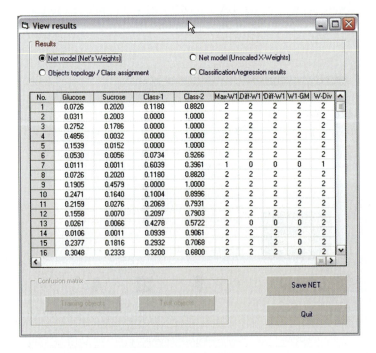

Figure 2.113. Milano

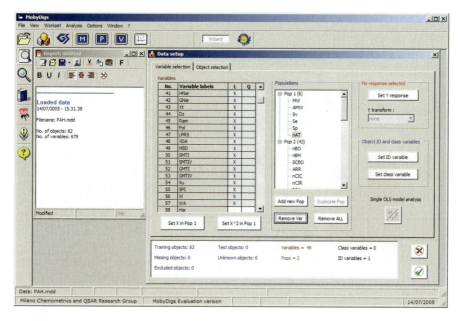

Figure 2.114. Milano

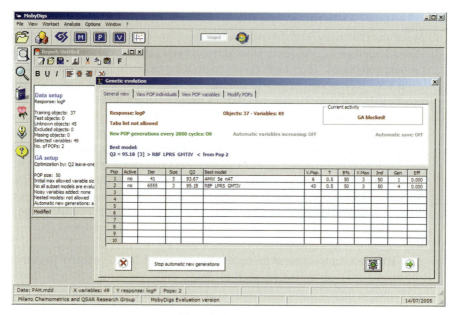

Figure 2.115. Milano

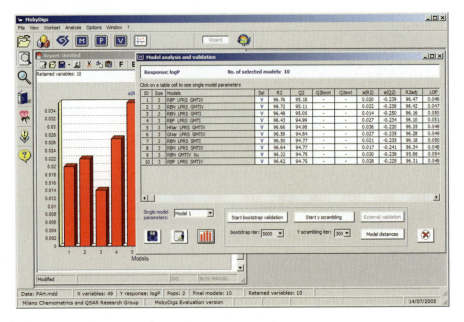

Figure 2.116.   Milano

2.25   MOLECULAR DISCOVERY

I. Molecular Discovery (http://www.moldiscovery.com/index.php)

II. Product Summaries: Molecular Discovery develops software tailored for pharmaceutical companies. Their tools enable researchers to derive high quality macromolecular 3D-descriptors and to manage the data explosion in order to link Structure-Based Drug Design, eADMET, Chemoinformatics and Bioinformatics. Their products merge molecular modeling techniques and QSPR methods into 3D-QSPR models to create a package whose application is drug discovery, property design and prediction. These strategies can be applied to virtual screening and prediction of human drug metabolism. The pipeline of integrated software includes several programs: GRID, VolSurf, Almond, MetaSite, and Penquins. Molecular Discovery is linked with Sunset Molecular Discovery LLC whose focus is computer-assisted lead discovery and understanding chemistry/biology interactions.

III. Key capabilities and offerings:

  a. **GRID** is a computational procedure for determining energetically favorable binding sites on molecules such as drugs, molecular arrays such as membranes or crystals, and macromolecules such as proteins, nucleic acids, glycoproteins or polysaccharides. GRID is useful in understanding the structural differences related to enzyme selectivity, a fundamental field in the rational design of drugs. GRID accelerates and simplifies the identification of trapped water molecules in hydrophobic regions between ligand and receptor. This information is useful in determining the necessary structural ligand modifications to lower the destabilizing effect between ligand and receptor. GRID maps can also be used as descriptors input in statistical procedures like CoMFA, GOLPE or SIMCA for QSAR or 3D-QSAR analyses.

b. **VolSurf** is a computational procedure to produce 2D molecular descriptors from 3D molecular interaction energy grid maps. VolSurf uses multivariate statistics coupled with interactive 2D and 3D plots to obtain valuable insights for drug design, PK profiling and screening. VolSurf compresses the information present in 3D maps into 2D numerical descriptors. VolSurf descriptors are specifically designed for the optimization of *in silico* pharmacokinetic properties (eADME or IS-DMPK). VolSurf working examples for passive intestinal and brain permeability, solubility, and cytochrome P450 inhibition are reported in the most recent literature.

c. **Almond** is a program specifically developed for generating and handling alignment independent descriptors called GRIND (GRid INdependent Descriptors). This new generation of 3D-molecular descriptors has applications in 3D-QSAR, QSAR, virtual screening and design of combinatorial libraries. ALMOND includes a complete package of chemometric tools adapted to the specific requirements of the new variables. With ALMOND software the following task are possible: biomolecule characterization, binding site studies, selectivity studies, quantitative structure-metabolism relationships and quantitative structure-transport relationships.

d. **MetaSite** is a computational procedure specially designed to predict the site of metabolism for xenobiotics starting from the 3D structure of a compound. This methodology has been developed to predict the site of metabolism for substrates of several cytochromes. The computations are completely automated and do not require any user assistance.

e. **Penguins**, Pharmacokinetics Evaluation aNd Grid Utilization IN Silico, is a cheminformatic tool for guiding drug discovery and development. It combines *in silico de-novo* chemical synthesis, biological screening and datamining approaches that allow for the rational selection of designed compounds with optimal pharmacokinetic and pharmacodynamic properties from an almost infinite number of synthetic possibilities.

IV. Review:

Molecular Discovery provides a versatile software package whose module capabilities range from helping researchers identify possible structural modifications that could improve ligand/receptor binding (GRID) to allowing for the rational selection of designed compounds with optimal pharmacokinetic and pharmacodynamic properties (Penguins). Molecular Discovery, in association with Sunset Molecular Discovery LLC, also provides consulting services for lead discovery.

Figure 2.117.   Molecular Discovery

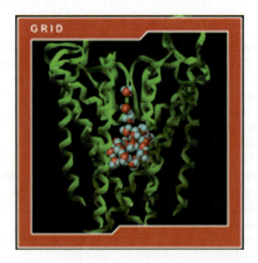

Figure 2.118.   Molecular Discovery

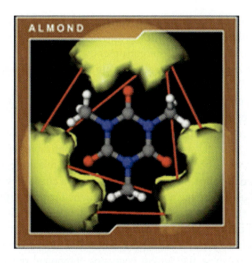

Figure 2.119. Molecular Discovery

2.26 MOLECULAR NETWORKS

I. Molecular Networks GmbH: (http://www.mol-net.com)
II. Product Summaries:

   a. *2DCOOR* coordinates generator for publishing quality 2D depictions.

   b. *ADRIANA* provides algorithms for the search, identification, and optimization of hits and lead structures.

   c. *ADRIANA.Code* calculation of molecular physicochemical properties, autocorrelation of 2D and 3D interatomic distance distributions, RDF of 3D interatomic distances and autocorrelation of distances between surface points.

   d. BioPath database that provides access to biological transformations and regulations as described on the Roche Applied Science "Biochemical Pathways" wall chart.

   e. *C@ROL* data warehousing for 2D structures, multiple 3D conformations, and experimental information.

   f. *CHECK* structure integrity check and normalization of chemical state.

   g. *CONVERT* inter-conversion of 40 different chemical file formats.

   h. *CORINA* 3D structure generation.

   i. CORINA.*direct* graphical user interface for CORINA including a molecule editor and a 3D structure viewer.

   j. *CORINA_F* CORINA interfaced to FlexX docking program.

   k. *IMAGE* conversion of chemical files into images.

   l. *PAGE* conversion of chemical files into formatted documents.

   m. *ROTATE* generation of ensemble of conformations.

   n. *SONNIA* self-organizing neural network package.

o. *SPLIT*/JOIN&MERGE splitting and concatenating of a chemical file or merging with external data files.

p. *STERGEN* enumeration of stereoisomers.

q. *TABLE* conversion of chemical files into spreadsheet file formats.

r. *TAUTOMER* enumeration of tautomers.

s. *WODCA* synthesis design by retro-synthetic analysis.

III. Key capabilities and offerings:

a. **Warehousing Structures and Data:** *C@ROL* is a chemical warehouse system designed to store 2D structures, multiple 3D conformations of chemical compounds as well as chemical reactions along with related (e.g., experimental or computed) data. The system is implemented as a client-server application. The web-based user interface provides access to the features of the structure search engine for the retrieval of chemical compounds and their related data. This engine can perform structure and sub-structure search, similarity search, and transformation search. When C@ROL is linked to the Commercially Available Compound database the system generates compound purchase orders. There are a variety of methods for Structure, Reaction and Data Retrieval:

   i. String Search

   ii. Property Search

   iii. Full Structure and Sub-Structure Search

   iv. 3D Pharmacophore-Type Search

   v. Reaction Center Search

   [Reitz M, Sacher O, Tarkhov A, Trümbach D, Gasteiger J, (2004) Enabling the exploration of biochemical pathways. Org. Biomol. Chem. *2*, 3226–3237.]

b. **Generating 2D Coordinates:** 2DCOOR generates high-quality 2D depictions (atomic coordinates) of chemical compounds. 2D depictions are 2-dimensional representations of chemical compounds similar to the ones a chemist would sketch. 2DCOOR is able to align the layout of images according to a given substructure template

c. **Generating 3D Coordinates:** *CORINA* generates 3-dimensional molecular models information on atom types and atom connectivity only. CORINA is used routinely for conversion of large datasets. This application is currently used by MDL, NIH/NCI and all major pharmaceutical companies to convert their 2D structures into 3D. Although often represented in 2D (2D Depictions) by chemists, the molecular structure of a compound is three-dimensional (3D). This 3-dimensional structure is closely associated with the chemical, physical, and biological properties of chemical compounds. 3D structures are starting materials for 3D QSAR, ligand-protein docking and drug design studies. CORINA can be used in batch mode or interactively by applying the graphical user interface CORINA.*direct*. *CORINA_F* is a feature-restricted version of CORINA that has been interfaced to FlexX, the flexible docking program distributed by BioSolveIT GmbH. [Sadowski J, Gasteiger J, (1993) From Atoms and Bonds to Three-dimensional Atomic Coordinates: Automatic Model Builders. Chem. Rev. 93: 2567–2581.]

d. **Controlling Structural State & Integrity:** *CHECK* performs high throughput structure integrity checks and can be used to normalize the state of each compound by applying certain business rules. Integrity check is performed on the atomic valence, hybridization state, and ionization state of a molecule. The state of each structure can be controlled by specifying its ionization state, by removing salts and solvents or by adding missing hydrogen atoms. Furthermore, in large structure files duplicate structures can be detected and removed.

e. **Enumerating Stereoisomers & Tautomers:** *STERGEN* automatically identifies stereocenters (tetrahedral centers and cis/trans double bonds) and enumerates all possible combinations of stereoisomers. This application can be used to process large datasets of chemical structures. *TAUTOMER* enumerates all tautomeric forms of a compound. TAUTOMER can be restrained to generate only one tautomer, the proposed form being assumed to be one of the most prevalent in solution.

f. **Exploring Conformational Space:** *ROTATE* automatically generates conformational ensembles from a starting 3D structure like the one obtained with CORINA, including conformers, which come close to biologically active ones. ROTATE can perform a user-defined and balanced sampling of the conformational space in order to obtain a sets of diverse conformations. [Schwab, C.H. (2003) Conformational analysis and searching. In: Handbook of chemoinformatics – from data to knowledge. Gasteiger J (ed.) Weinheim D, Wiley-VCH. pp. 262–301.]

g. **Computing Descriptors:** *ADRIANA.Code* calculates a series of molecular descriptors that can be applied in the area of in silico discovery and optimization of new chemical entities. The descriptors encode physicochemical, topological, geometrical, and surface properties of molecules [Gasteiger J (2003) Physicochemical effects in the representation of molecular structures for drug designing. Mini Rev Med Chem 3: 789–796.]. The following descriptors can be calculated:

   i. Global molecular descriptors including number of H bond acceptors and H bond donors, TPSA, molecular weight, dipole moment, log P, log S and molecular polarizability

   ii. Autocorrelation of 2D and 3D interatomic distance distributions weighted by partial charges, electronegativities, and polarizabilities [Gasteiger J, Teckentrup A, Terfloth L, Spycher S (2003) Neural networks as data mining tools in drug design. J Phys Org *Chem 16*: 232–245.]

   iii. Radial Distribution Functions of 3D interatomic distances weighted by partial charges, electronegativities, and polarizabilities [Terfloth L, Gasteiger J (2003) Electronic screening: lead finding *from database mining*. In: The practice of medicinal chemistry, 2nd Edition, Wermuth CG (ed.) Amsterdam, NL, 2003, pp. 131–145.]

   iv. Autocorrelation Functions of distances between surface points weighted by molecular electrostatic potential, hydrogen bonding potential, and hydrophobicity potential [Teckentrup A, Briem H, Gasteiger J (2004) Mining high-throughput screening data of combinatorial libraries: development of a filter to distinguish hits from nonhits. J. Chem. Inf. Comput. Sci. 44: 626–634.]

h. **Analyzing and Modeling:** *ADRIANA* (Automated Drug Research by Interactive Application of Non-linear Algorithms) bundles the two software packages *ADRIANA.Code* and *SONNIA*. *SONNIA* is a self-organizing neural network package including both unsupervised (Kohonen) and supervised (counter-propagation network) learning techniques. SONNIA has a graphical user-interface for the visualization of chemical structures, reactions, and spectra. Statistical or machine learning methods are widely used to establish relationships between biological activities, physical or chemical properties of a compound and its chemical structure. These methods, in combination with structure descriptors, are used to derive models that can be applied to predict properties of new compounds. [Anzali S, Gasteiger J, Holzgrabe U, Polanski J, Sadowski J, Teckentrup A, Wagener M (1998) The use of self-organizing neural networks in drug design. In: Kubinyi H, Folkers G, Martin YC (eds) 3D QSAR in Drug Design – Volume 2, Kluwer/ESCOM, Dordrecht, NL, pp. 273–299.]

i. **Warehousing Reactions:** *C@ROL* is a molecule-oriented data warehousing system for the storage of chemical reactions, 2D structures, and multiple 3D conformations of chemical compounds. C@ROL is a client-server application that includes a structure search engine for the retrieval of chemical reactions and their related data. This engine can perform reaction searches, structure and sub-structure searches, similarity searches and precursor searches.

j. **Designing Synthesis:** *WODCA* provides synthesis design solutions for organic compounds. This application relies on a retrosynthetic approach to provide acceptable synthetic pathways. By identifying strategic bonds in the product, WODCA suggests suitable precursors. Retrieval in reaction databases shows the viability of retrosynthetic steps and provides experimental conditions. This procedure can be applied recursively until commercially available starting materials are identified. [Gasteiger J (2003) The prediction of chemical reactions. In: Gasteiger J, Engel T (eds), Chemoinformatics – A Textbook, Wiley-VCH, Weinheim, pp. 542–567. Sitzmann M, Pförtner M (2003) Computer-assisted synthesis design. In: Gasteiger J, Engel T (eds), Chemoinformatics – A Textbook, Wiley-VCH, Weinheim, pp. 567–593.]

k. **Converting and Manipulating files:** *CONVERT* enables the inter-conversion of 40 different structure and reaction file formats. This application automatically detects the format of the input file and converts it into the format specified by the user. *TABLE* converts files containing structures and data into spreadsheet and internet compatible file formats. Formats supported include Excel, dBASEIII and HTML. During that process, structures are converted to WMF (Excel) or GIF images. SPLIT/JOIN&MERGE splits a structure file including *n* structures into *n* separated files, concatenates a series of *n* structure files into one single file, or merges separated structure files and data files into a single SDFile.

l. **Drawing and printing:** *IMAGE* is designed to convert structure files into raster (GIF, PNG, BMP) or vector (WMF, EMF, EPS) images. These file formats can be used to embed 2D depicts or flattened 3D structures into documents. These file formats can also be imported in digital imaging software for graphic design. *PAGE* converts chemical files into formatted documents. The application provides many parameters to control the page layout. The resulting PostScript file can printed or converted into a PDF document.

m. **Databases:**

  i. *C@ROL* database includes about 1 million compounds from diverse commercial providers. These structures are stored with physicochemical properties like molecular weight, predicted log P, solubility and Lipinsky rule of 5. Structure, substructure and similarity searches can be performed along with any regular data and string search.

  ii. The Biochemical Pathways database (BioPath) contains data derived from the *Roche Applied Science* "Biochemical Pathways" wall chart. BioPath provides access to biological transformations and regulations as described on the "Biochemical Pathways" chart. [Reitz M, Sacher O, Tarkhov A, Trümbach D, Gasteiger J (2004) Enabling the exploration of biochemical pathways. Org Biomol Chem *2*, 3226–3237.]

iv. Review:

Molecular Networks products have been largely developed by the work by Professor Gasteiger on a range of chemoinformatics technologies to depict molecules and chemical reactions as described above. Many of these programs have been widely used, of particular note is the CORINA program which is broadly used to generate 3D conformations of 2D renditions of structures and integrated into other vendor products. Molecular Networks collaborates with MDL for databases that feature CORINA-generated models to provide researchers with a more complete and realistic set of searchable models in an industry-standard format. Molecular Networks collaborates with SciTegic to integrate CORINA and other programs into Pipeline Pilot data processing protocols. Molecular Networks collaborates with Biomax to join bioinformatics and chemoinformatics into one stream of research, with BioSolveIT to integrate collections of virtual screening tools to facilitate *in silico* synthesis of chemical fragments optimized towards certain target-dependent properties (see CORINA_F), and with Inte:Ligand to integrate CORINA into i:lib diverse for virtual compound library generation.

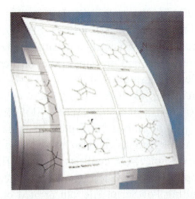

Figure 2.120.   Molecular Networks

Figure 2.121.   Molecular Networks

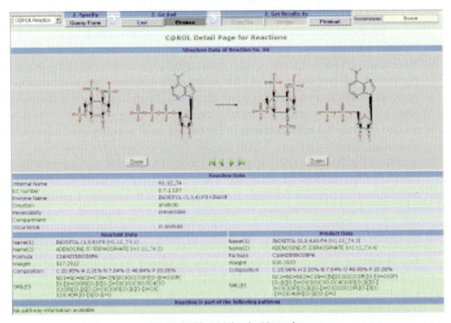

Figure 2.122.   Molecular Networks

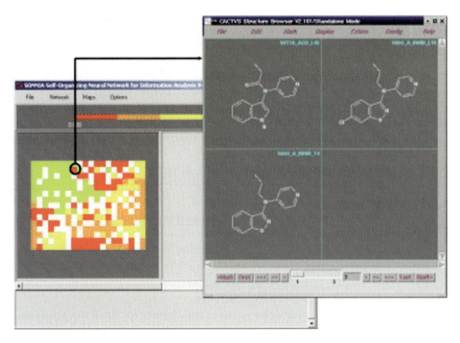

Figure 2.123. Molecular Networks

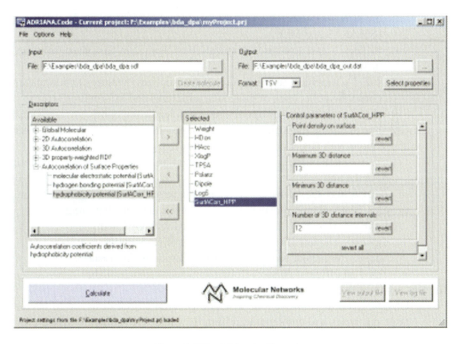

Figure 2.124. Molecular Networks

2.27   OPEN EYE SCIENTIFIC SOFTWARE

I. Open Eye Scientific Software; http://www.eyesopen.com/
II. Product Summaries:
    a. General: OpenEye Scientific Software, Inc. develops large-scale molecular modeling applications and toolkits. Primarily geared towards drug discovery and design, areas of application include structure generation, docking, shape comparison, charge/electrostatics, chemical informatics and visualization. OpenEye makes much of its technology available as toolkits – programming libraries suitable for custom development.
    b. OpenEye offers the following application software:
        i. *EON* – chemical similarity analysis via comparison of electrostatics overlay.
        ii. *FILTER* – molecular screening and selection based on physical property or functional group.
        iii. *FRED* – fast, systematic docking search for ligand binding within a protein active site.
        iv. *OMEGA* – systematic high-throughput conformer generation, including 1D or 2D to 3D structure generation.
        v. *QUACPAC* – quality charge states and charges for small molecules and proteins.
        vi. *ROCS* – chemical similarity analysis via rapid 3D molecular shape searches.
        vii. *SMACK* – molecular databases query converter and optimizer (SMARTS and MDL).
        viii. *SZYBKI* – fast structure optimization of ligands in gas-phase, solution, or within a protein active site.
        ix. *VIDA* – graphical user interface that visualizes, analyzes and manages corporate collections of molecular structures and information.
        x. *WABE* – electrostatics optimization of a lead compound.
    c. OpenEye offers the following toolkits as programming libraries providing other applications with object-oriented accessibility to a given set of capabilities:
        i. *Case* – generalized function optimization, e.g. molecular structure optimization.
        ii. *LexiChem* – state-of-the-art compound name and structure interconversion.
        iii. *OEChem* – chemoinformatics and 3D molecular data handling.
            1. *SCUT Monkeys* – all the apps included with OEChem.
        iv. *Ogham* – elegant 2D structure rendering of compounds.
        v. *Shape* – molecular shape comparisons based on 3D overlays.
        vi. *Zap* – an efficient Poisson-Boltzmann electrostatics solver.
III. Key capabilities and offerings:
    a. OMEGA is a high-throughput structure generation tool designed to handle large databases and combinatorial libraries valuable to computer-aided drug

design. Exhaustive conformational expansion of drug-like molecules can be performed in fractions of a second using a systematic algorithm to ensure reproducibility, yielding a throughput of hundreds of thousands of compounds per processor per day. Omega accepts a wide-variety of input file formats, including 1D connection tables. Conformers can be stored in a number of formats, including an ultra-compact one averaging 20 bytes per structure. The output ensembles are designed to include the bioactive conformer and can be minimized against MMFF and solvent forces. OMEGA provides a natural entry point to OpenEye structural analysis software and is available for a variety of operating systems and hardware. Over a dozen features are described on the website.

b. ROCS is a shape comparison program, based on the idea that molecules have similar shape if their volumes overlay well. Any volume mismatch is a measure of dissimilarity. Implementation of the global search required to find the best overlay is difficult with hard-sphere representations. Instead ROCS uses a Gaussian representation of the molecular volume. Because the volume function is smooth, it is possible to routinely minimize to the best global match from a few starting arrangements. ROCS is capable of processing 600–800 queries each second. At this speed it is possible to search multi-conformer representations of corporate collections in a day on a single processor. ROCS is also able to include chemical knowledge by the addition of a SMARTS-defined chemical force field. The force field can be discrete or Gaussian. Discrete scores the overlays based on proximity of SMARTS patterns, while Gaussian weights such scores based on distance. Furthermore, because the Gaussian force field is differentiable, it can be included in the minimization to arrive at a different overlay than would be found from shape alone. Nine features are described on the website.

c. OEChem is a programming library for chemistry and chemical informatics that is fast, stable and well documented. OEChem has many simple yet powerful functions, which handle the details of working with molecules. For routine tasks, OEChem offers clear and efficient scripting in Python; over 70 scripts are provided as examples for common tasks. For more advanced software development and enterprise solutions, OEChem offers a stable API in Python and C++. High-level functions provide simplicity while low-level functions provide flexibility. OEChem is available on many platforms and is the core chemistry toolkit for all OpenEye products. Fifteen key features of the OEChem Toolkit are described on the website.

d. FILTER is a molecular screening and selection tool that uses a combination of physical-property calculations and functional-group knowledge to assess compound collections. In selection mode, FILTER can be used to choose reagents appropriate for specific syntheses. In filter mode, it quickly removes compounds with undesirable elements, functional groups, or physical properties. FILTER is a command line utility that, like other OpenEye products, reads and writes numerous file formats. FILTER currently has the ability to select or reject compounds based on eleven criteria described on the website, including extensive definitions of undesireable functional groups.

e. ZAP: Poisson-Boltzmann (PB) is an efficient way to simulate electrostatics in a medium of varying dielectric, such as organic molecules (drugs, proteins) in water. It requires a molecular charge description and a designation of low (molecular) and high (solvent) dielectric regions. The ZAP toolkit provides facilities to produce a grid of PB electrostatic potentials and, from this, a long list of biologically interesting quantities. These include solvent transfer energies, binding energies, pKa shifts, solvent forces, electrostatic descriptors, surface potentials and effective dielectric constants. Unique to ZAP is a dielectric function based on atomic-centered Gaussian functions. ZAP avoids many pitfalls of discrete dielectric models and works well not only for small molecules but also proteins and macromolecular ensembles. Eight key features of the ZAP toolkit are described on the website.

f. VIDA is a graphical interface designed from the ground-up to browse, manage and manipulate large sets of molecular information. Built-in chemoinformatics (SMARTS-matching, SMILES parsing), an advanced list manager, spreadsheet, annotation and graphing capabilities make it possible to operate in real-time on corporate collections of a million structures. It supports all standard visualization paradigms for both small molecules and proteins, including 2D depictions, both hardware and software stereo, and many unique facilities including surface selection and manipulation. Many methods of physical property calculation are included, and a wide range of formats can be read and written. VIDA can also act as an interface to other OpenEye's modeling software for initial setup or analysis of generated data. The eighteen primary features of VIDA are described on the website.

g. FRED stands for Fast Rigid Exhaustive Docking. For every ligand, FRED exhaustively searches all possible poses within a protein active site, filtering for shape complementarity and pharmacophoric features before evaluating with several scoring functions (ChemScore, PLP, ScreenScore, ChemGauss, PBSA). FRED uses a systematic search algorithm, accurately predicting binding modes in a reproducible manner, unlike many other docking programs, which use stochastic methods. Despite being exhaustive, FRED is extremely fast, out-performing all competitive methods; FRED docks about a dozen ligand conformers per second per processor. Furthermore, FRED will perform constrained docking, wherein certain pharmacophoric features are guaranteed to be in specific regions of the active site, allowing scientists to take advantage of known structure activity relationships. The twelve primary features of FRED are described on the website.

h. SHAPE is an object-oriented toolkit in C, modeled in form after the Daylight chemical toolkit. SHAPE makes accessible the functions and capabilities that underlie ROCS, OpenEye's shape comparison program, and facilitates the incorporation of molecular overlay into other applications. SHAPE encompasses the calculation of molecular descriptors for shape (steric multi-poles), the volume overlap between molecules, the spatial similarity of chemical groups (color force field) and the optimization of the latter two quantities. SHAPE utilizes a variety of methods, some tuned for performance, others for accuracy. Example applications are provided. SHAPE has been extended to allow operation on a generic shape fields, e.g. grids, and can be applied to such diverse problems as comparing and aligning active sites and real-space fitting of electron-density. Eight example uses of SHAPE are included on the website.

IV. Review:

OpenEye's modular software applications and toolkits allow developers to build their own tools. Open Eyes software is widely used by developers for creating customized technologies. Instead of the tools that traditionally dominate the field, both in academia and in industry, OpenEye attempts to provide tools that vastly increase the scale of operation of computational chemistry in drug design. OpenEye sells software designed to be useful in drug discovery and drug optimization. Simple products for non-programmers and non-experts could widen the benefit to larger communities of scientists. Java interfaces to all toolkits are currently under development to specifically meet this need.

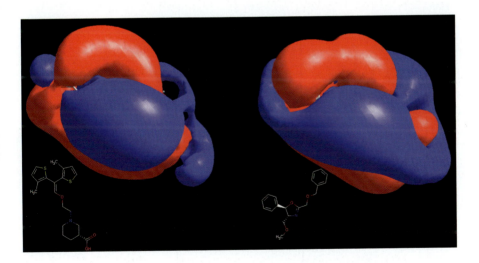

Figure 2.125. (eon_mdd)   Two molecules from the MDDR with substantially different chemistry, but high shape and electrostatic similarity ($T_{shape} > 0.75$, $T_{electrostatic} > 0.3$)

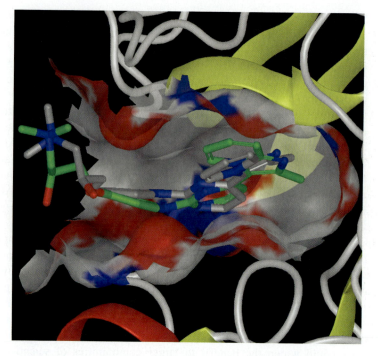

Figure 2.126. (fred_cdk2)   The crystallographic structure of a CDK-2 ligand
in its bound conformation (green) overlaid by the pose predicted by FRED

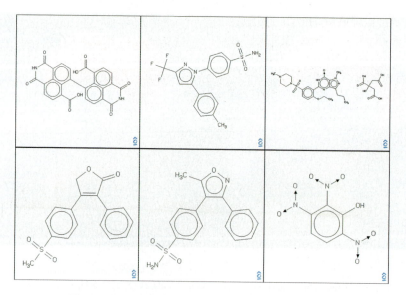

Figure 2.127. (ogham_land)   2D depictions for six compounds

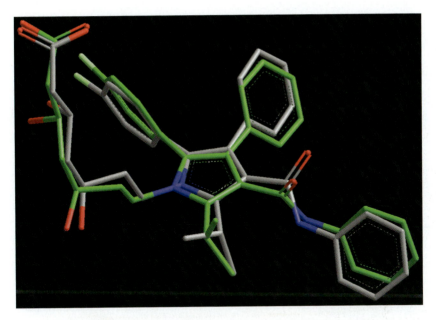

Figure 2.128. (omega_lipitor)   The Xray crystal structure of Lipitor in its active conformation (green) overlaid by an Omega conformation which reproduces the active conformation to within 0.75 Å rms

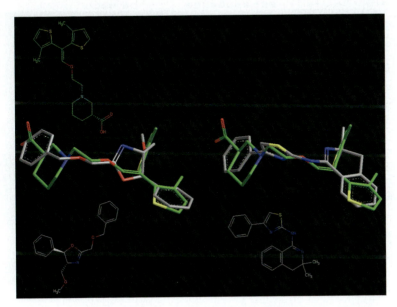

Figure 2.129. (rocs_similarity)   Two molecules with substantially different chemistry from the query (green), but high shape similarity ($T_{shape} > 0.75$)

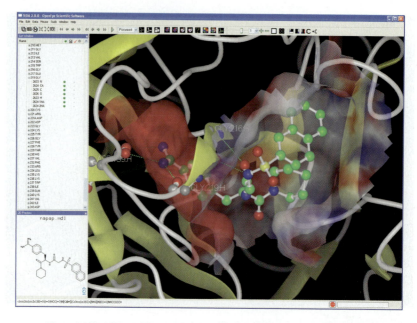

Figure 2.130. (vida_docking)   Visual inspection of the interactions of a ligand,
napap, in the active site of a protein, thrombin

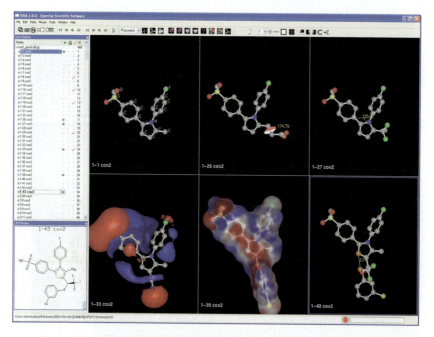

Figure 2.131. (vida_multipane)   Visual inspection of a collection of small molecules,
here cox2 inhibitors, is straightforward using a tiled view

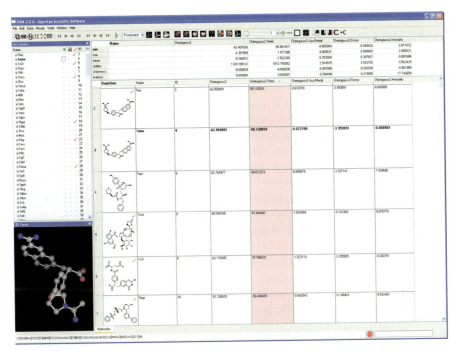

Figure 2.132. (vida_spreadsheet)   A spreadsheet facilitates data analysis, such as
sorting and annotating results from a docking run

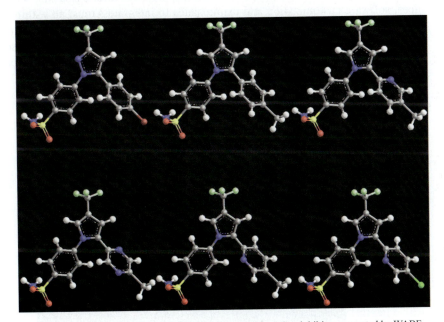

Figure 2.133. (wabe_6cox)   Several isosteric analogs of a 6 Cox inhibitor generated by WABE

2.28   PLANARIA-SOFTWARE

I. Planaria-Software (http://www.planaria-software.com/)
II. Product Summaries: Planaria distributes the ArgusLab molecular modeling program.
III. Key capabilities and offerings:
   a. ArgusLab is a complete modeling package that includes a 3D builder/visualizer and supports a wide array of quantum and molecular mechanics calculations. Also included is ArgusDock, a new drug docking code which includes Genetic Algorithm and TreePruning docking engines and the AScore scoring function. The docking code supports interactive docking and docking of ligand databases. Other features include an interface to running Gaussian calculations on a Windows PC in a fully integrated manner, a treeview tool for navigating and editing structures and calculation results, several enhancements to ZINDO that include triplet excited states and electric field calculations, a peptide builder that generates alpha helices and beta strands, depth cueing, ribbon rendering of proteins, integrated downloading of structure files from the PDB, solvent-accessible surfaces, and improved surface rendering quality. The molecule builder also allows users to export images to bitmap for graphic presentations. Features such as builder toolkit and quick plot buttons are helpful in creating HOMO-LUMO and ESP-mapped density surfaces. An added graphical feature is the ability to export image files to the POV-Ray ray-tracer (freely available at http://www.povray.org).
IV. Review:
   ArgusLab's graphical abilities and breadth of calculations are the real pluses of this intuitive, easy to use software. In addition, they also offer consulting services, software customization and maintenance to integrate ArgusLab within a client's corporate environment. The integrated HTML help docu'mentation is good, but somewhat incomplete on the newer features. The authors promise updated documentation. There is an active users group that is useful for tips and asking questions (http://www.arguslab.com/ usersgroup.htm).

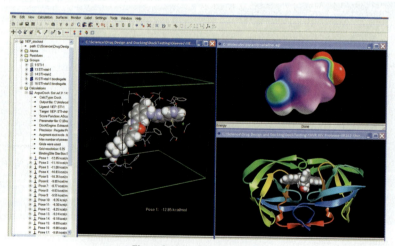

Figure 2.134.   Planaria-Argus

2.29   PUBCHEM

I. Pubchem; http://pubchem.ncbi.nlm.nih.gov/
II. Product Summaries:
    a. Pubchem is an NIH funded effort that was initiated in 2004 to provide chemically annotated information for free to the scientific community. The current summary is as of Q2/2005.
    b. PubChem contains the chemical structures of small organic molecules and information on their biological activities.
    c. PubChem is organized as three linked databases within the *Entrez/PubMed* information retrieval system. These are PubChem Substance, PubChem Compound, and PubChem BioAssay:
        i. ***PubChem Compound***: Search for unique chemical structures using text terms such as names, synonyms, keywords, or depositor-supplied identifiers.
        ii. ***PubChem Substance***: Search for deposited structure records using text terms such as names, synonyms, keywords, or depositor-supplied identifiers.
        iii. ***PubChem BioAssay***: Search for bioassay records using text terms from the bioassay description, such as the names of reagents used in the bioassay.
        iv. ***PubChem Structure Search***: Search PubChem/Compound using chemical structure. Structure may be specified using SMILES, MOL file, and other formats.
        v. PubChem also provides PubChem Structure Search, a chemical structure similarity search tool that links to the PubChem Compound and PubChem Substance databases.
        vi. A *PubChem FTP* site is also available. The Pubchem FTP site has files in ASN, XML, SDF, and CSV formats.
III. Key Capabilities and Offerings:
    a. PubChem is NIH's Small Molecule Repository which is a component of the National Institutes of Health (NIH) Roadmap Molecular Libraries initiative for the creation of an NIH small molecule library and screening centers.
    b. PubChem is integrated with *Entrez*, NCBI's primary search engine, and also provides compound neighboring, sub/superstructure, similarity structure, bioactivity data, and other searching features.
    c. A Pubchem Help page provides a general overview of capabilities: http://pubchem.ncbi.nlm.nih.gov/help.html. The document provides tips and examples for searches of the three PubChem databases by text term/keyword, as well as tips for searching PubChem Compound by chemical properties. Additional help documents provide tips on using chemical information for *basic* and *advanced* structure search options in the *PubChem Structure Search*.
    d. Advanced structure search allows one to search structures, substructures, and similiary to draw structures by clicking on edit to generate a structure drawing palette using the Cactvs server side structure editor.

e. PubChem substance database contains chemical structures, synonyms, registration IDs, description, related urls, database cross-reference links to PubMed, protein 3D structures, and biological screening results. If the contents of a chemical sample are known, the description includes links to *PubChem Compound*.

f. The *PubChem* Compound Database contains validated chemical depiction information provided to describe substances in *PubChem Substance*. Structures stored within PubChem Compound are pre-clustered and cross-referenced by identity and similarity groups. Additionally, calculated properties and descriptors are available for searching and filtering of chemical structures. Examples of chemical property search are also provided.

g. The *PubChem* BioAssay Database contains bioactivity screens of chemical substances described in *PubChem Substances*. It provides searchable descriptions of each bioassay, including descriptions of the conditions and readouts specific to that screening procedure.

h. The PubChem's default results page is part of the Entrez summary list display system.

    i. *PubChem Substance results*: From Entrez PubChem substance database, users can get substance summary with thumbnails, corresponding compound ID, depositors source information, etc

    ii. Users can choose to display brief, summary, ID map, substance neighboring information, synonyms, and other database's information from the dropdown list. From the right side pop-up window, users can select links related to this substance, such as PubMed, Mesh, BioAssay, neighboring substances, etc. Users can choose to either '*display*', or '*send*' the searched results to '*text*' or a '*file*'. Users can reach the more detailed substance information and cross links by click the structure image or the id link.

    iii. The substance's standardized compound information also lets users get property data, synonyms, descriptors, comments, cross links, depositor's structure drawing, etc. Power users even can download different data formats, such as ASN.1, XML, and SDF, for further use.

    iv. With the *compound summary page* users can find this compound's property data, description, related substance information, neighboring structures, and cross links.

    v. All compounds are structurally unique by comparing each other. So one compound may link to may substances.

i. BioAssay result maybe navigated using the PubChem BioAssay assay browser. You may select descriptors to view in an assay in the *BioAssay Summary* page. The *Search BioAssay Results* page provides an interface to specify search criteria for the assay descriptors. You may also combine the current query with other Entrez PubChem Substance searches. The query result counts will be shown in the *BioAssay Results Preview* page, where you may also select to display other chemical properties of the search result. The *BioAssay Results* page tabulated the search result with thumbnail structures, PubChem Substance IDs and other assay and compound descriptors specified in your selections.

i. Using the "Display" menu in this page, users may choose to view lists of summaries, brief summaries, unique identifiers, compounds, substances, free text article links (via PMC) and PubMed citations. Similar to other Entrez databases, you may add the result to Entrez Clipboard, view as HTML text and save to text file using the "Send to" menu. The "Links" pull-down menu on the far right provides short-cuts to PubChem Substance and PubChem Compound screened in the BioAssay.

j. PubChem provides cross links to other databases when those information are available. You can find those links from either entrez PubChem pages or individual record summary pages.

   i. **SID:** Link to Entrez PCSubstance.

   ii. **CID:** Link to Entrez PCCompound.

   iii. **Neutral Form:** Link to Entrez PCCompound with this structure's neutralized form.

   iv. **Related Compounds and Substances Links:** on compound and substance summary pages respectively.

   v. **Similar Compounds and Substances Links:** All compounds shown have the similarity score over 90% [Tanimoto]. If you want to find compounds with different scores, you can visit the PubChem *structure search page*.

   vi. **Structure Search:** Lead you the PubChem structure search page by transferring this compound's isomeric SMILES string into the search field. For more information about the structure search, visit *structure search help*.

   vii. **BioActivity:** Link to bioassay data summary page.

   viii. **Structure Links:** Link to Entrez Structure database with associated mmdb ids provided by depositors.

   ix. **PubMed:** Links to PubMed references with this compound/ substance.

   x. **Nucleotides:** Linking to Entrez nucleotides database with associated nucleotides gis provided by depositors.

   xi. **Protein:** Links to Entrez protein database with associated protein gis provided by depositors.

   xii. **Source and Source-ID:** Links to depositor's original information page by depositor's source name and/or source-id (external id) if available.

   xiii. **Medical Subject Annotations (MeSH):** Linking information to records in NLM's Medical Subject Heading (MeSH) database. Linkage is based on matching names and synonyms supplied with the chemical structure record to those in the MeSH record.

   xiv. **NLM Toxicology Link:** Links to the SIS/ChemIDPlus record for this compound/substance, which provides further links to toxicology information sources.

k. In addition to PubChem, the National Library of Medicine also has developed ChemIDplus at http://sis.nlm.nih.gov/Chem/ChemMain.html for a database of over 368,000 chemical records, including over 240,000 with structures and 139,354 records with toxicity data. There are two versions of ChemIDplus which are part of the Specialized Information Services (SIS) Division of the National Library of Medicine (NLM) is responsible for information resources and services in toxicology, environmental health, chemistry, HIV/AIDS, and specialized topics in minority health:

    i. *ChemIDplus Lite* is available for Name and RN searching without the need for plugins or applets.

    ii. *ChemIDplus Advanced* allows for structure searching, plus biological and chemical property search and display.

IV. Review: Pubchem is integrated with the other Entrez/PubMed tools. Pubchem currently contains 850,000 compounds but will likely grow rapidly in size as websites with publicly available data begin to mirror content. It is by design destined to be the public repository for heterogeneous bioassay data generated by the NIH Intramural and Extramural screening centers. The NIH is also setting up a series of Exploratory Chemoinformatics Research Centers to evaluate how best to mine, search, model, and share data.

Figure 2.135. Pubchem

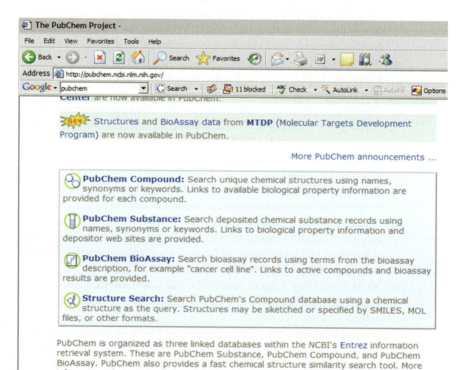

Figure 2.136. Pubchem

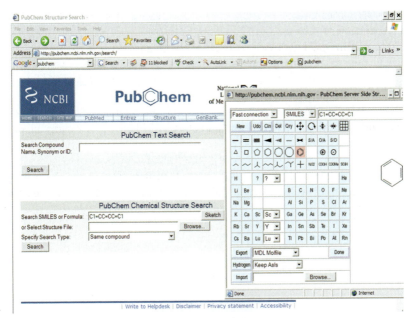

Figure 2.137.   Pubchem

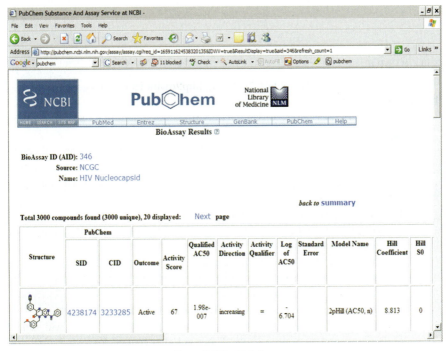

Figure 2.138.   Pubchem

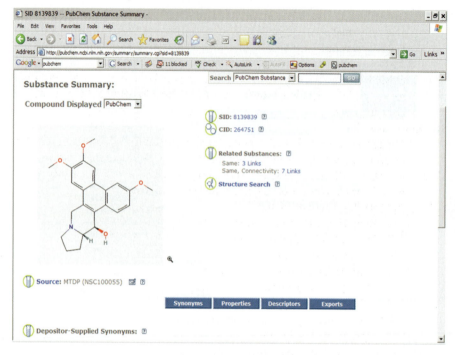

Figure 2.139. Pubchem

2.30 PYMOL

I. PyMOL; http://www.pymol.org; hosted at: http://pymol.sourceforge.net/

II. Product summary:

   a. PyMOL is a user-sponsored molecular modeling system with an open-source foundation.

   b. More specifically, PyMOL is a molecular graphics system with an embedded Python interpreter designed for real-time visualization and rapid generation of high-quality molecular graphics images and animations. It can also perform other tasks such as editing PDB files.

   c. PyMOL was designed for several things:

      i. visualizing multiple conformations of a single structure [trajectories or docked ligand ensembles]

      ii. interfacing with external programs,

      iii. providing professional strength graphics under both Windows and Unix,

      iv. preparing publication quality images, and

      v. fitting into a zero dollar budget (although support is requested).

   d. ***PyMOL 0.92*** has been released for *Windows, Linux, G4 Macs, G5 Macs, SGIs, Suns*, and as *Source Code*. There are also special *RedHat RPMS and Windows binaries* that will enable one to "import pymol" from external Python code in order to use PyMOL as a simple viewer module.

e. PyMOL was created to overcome limitations inherent in visualization pack-ages that make it difficult or impossible to get exactly what you need. PyMOL was designed open-source to enable users to surmount problems and innovate without unnecessary restraints.

f. PyMOL can be licensed for free and incorporated into other open or closed source projects. The intention is that others will add other layers and Graphic User Interfaces.

III. Key Capabilities:

a. The extensible core PyMOL module (hosted at SourceForge) is available free to everyone via the "Python" license (a simple BSD-like permission state-ment), but all users are asked to purchase a license and maintenance agree-ment in order to cover development and support costs.

b. PyMol's Top Features include:
   i. Real-Time 3D visualization.
   ii. Publication quality renderings.
   iii. Animation capabilities.
   iv. Support for X-ray crystallography.
   v. Modular architecture.
   vi. API for custom applications.
   vii. Open source.
   viii. Written in C and Python.

c. Multiple ways to control PyMol:
   i. Manually using internal or external GUI.
   ii. Manually using PyMol command language or from Python using the PyMol API.
   iii. Automatically using text format scripts written in the PyMol command language.
   iv. Automatically from Python programs which call the PyMol API or a combination of ways.

d. Though the program is Open-Source, it is best thought of as a modular semi-opaque tool, which can be extended an expanded through Python, rather than as a coding environment in which to embed new technologies. An extensive list of the strengths and weaknesses as of May 2003 are provided at: http://pymol.sourceforge.net/newman/user/S0107preface.html and the com-plete manual may be obtained for a fee.

IV. Review:

PyMOL is useful for viewing molecules and proteins. Because it is an open source architecture, in principal, it is much more scalable for custom applications than closed source software. Although the project has relatively a limited num-bers of applications at this early stage, it is envisioned to eventually become a full molecular modeling platform. Currently it is primarily a visualization package. Three main use cases are currently envisioned: first, to visualize molecule (sdfile) or protein (pdf file), second to create a publication quality figure (static content), and third to enable one to use Pymol as one does Powerpoint for gener-ating movies.

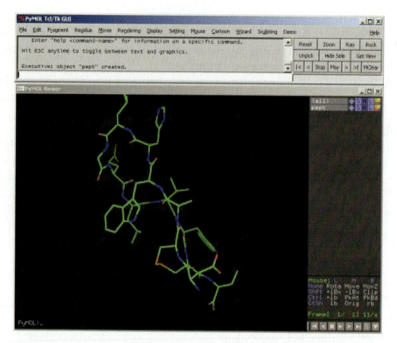

Figure 2.140.   PyMol

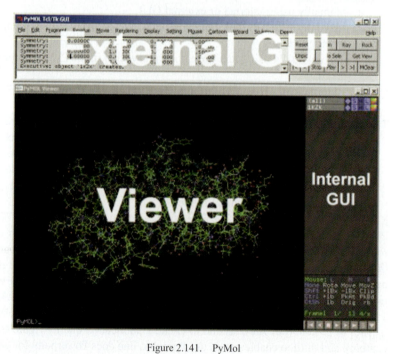

Figure 2.141.   PyMol

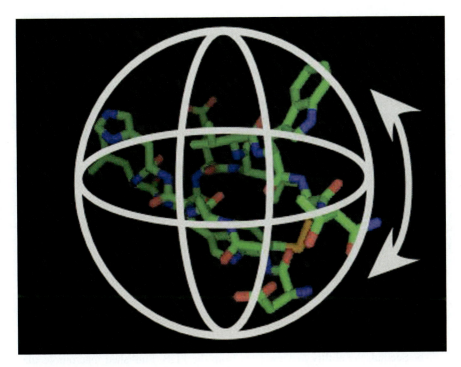

Figure 2.142. PyMol

2.31 RASMOL AND PROTEIN EXPLORER

I. RasMol and Protein Explorer; http://www.umass.edu/microbio/rasmol/ and http://molvis.sdsc.edu/protexpl/frntdoor.htm, respectively.

II. Product Summary:

a. *Protein Explorer*, a RasMol-derivative, is relatively easy to use and free software for looking at macromolecular structure and its relation to function. It runs on Windows or Macintosh/PC computers. It is very fast: rotating a protein or DNA molecule shows its 3D structure.

b. Ras Mol is *still available*, but it is much *harder to use effectively and considerably less powerful* than version 2 of *Protein Explorer* (released summer 2000). For individuals new to RasMol using Windows or Macintosh, it is strongly recommend that you start with **Protein Explorer** instead of RasMol. Protein Explorer (PE) is *much easier to use, and much more powerful*. PE's first image of a molecule is maximally informative (and explained) while RasMol's is an uninformative wireframe display (no explanation). PE has a "select" menu (lacking in RasMol), and you don't need to learn any commands to use PE. PE can show molecular surfaces (not available in RasMol), salt-bridges, and cation-pi interactions (each in one click). Every color scheme (e.g. polar/nonpolar, lacking in RasMol's Color menu) is accompanied by a color key.

c. Chime adds a great deal of power to RasMol, but it takes some time to fully utilize. The goal of Protein Explorer is to make the power of Chime accessible to students, educators, and scientists. More is available *about the evolution of the uses of Chime and the purpose of Protein Explorer.*

d. Version 2.0 of Protein Explorer, released in increments in 2000–2002, is much easier to use, and much more powerful than RasMol. Because Chime is available only for Windows and Macintosh, Protein Explorer runs best on those platforms; however, there are several methods for getting it to *work on linux, or other platforms.*

e. Protein Explorer is offered as unsupported freeware. All use is at the risk of the user. No warranty whatsoever is made or implied. This version is free for all users, but downloaded copies or derivatives thereof may not be publically redistributed on CD's, served from publically-accessible websites, or publically redistributed by other means without permission. Permission is given to link to Protein Explorer freely from other websites, provided no fee is charged for access.

III. Key Capabilities

a. First and foremost, the ability to view proteins, DNA, RNA and small molecules.

b. Second, the ability to view files from the PDB. Published macromolecular structures are archived at the Protein Data Bank (PDB). Search for the molecule you want, and then click on the *Protein Explorer* link built into the search result page. Or, notice the PDB code, and enter it.

c. Protein Explorer is free software for visualizing the three-dimensional structures of protein, DNA, and RNA macromolecules, and their interactions and binding of ligands, inhibitors, and drugs. It is suitable for high school and college students (ages 16 years and older), yet it is also widely used by graduate students and researchers.

   i. Protein Explorer enables one to see the relations of 3D molecular structure to function. The image can be simplified by hiding everything except the region of interest. A variety of one-click renderings and color schemes help to visualize the backbone, secondary structure, distributions of hydrophobic vs. hydrophilic residues, noncovalent bonding interactions, salt bridges, amino acid or nucleotide sequences, sequence-to-structure mappings and locations of residues of interest, and patterns of *evolution and conservation.*

   ii. Exploration is done largely from menus and buttons in Protein Explorer's QuickViews interface. The easy user interface means that you do not need to learn any "RasMol command language", although those who have learned some can use it freely.

   iii. Explanations, color keys, and help are displayed automatically with each operation you perform. Protein Explorer is a knowledge base with introductory information on many topics about protein structure, such as the origins and limitations of 3D protein structure data, specific oligomers vs. crystal contacts, hydrogen bonds, cation-pi interactions, etc. Protein Explorer includes an extensive *glossary and index*.

iv. One does not need to learn any *hand-typed commands* to achieve a very high level of visualization power. The *menus and buttons* in *PE's QuickViews* will do nearly everything.

v. There is a *1 Hour Tour* to get started, a *FAQ*, and a *Help/Index/Glossary* with over 200 entries to help figure out how to navigate. Key resources are introduced on Protein Explorer's *FrontDoor*.

vi. Educators have *lesson plans* and help with *assessment*, as well as an extensive illustrated *Atlas of Macromolecules*.

vii. PE has become an integrated knowledgebase:
1. Extensive help is displayed automatically for *every* menu operation. For example, if you COLOR ± *Charge*, you are told how and where to calculate the charge or pI of your protein on-line. If you SELECT *Nucleic*, there is a button to help you distinguish DNA from RNA.
2. PE has molecule-specific links to external resources, for example to visualize specific oligomers or virus capsids, or to color by evolutionary conservation or mutation.

viii. There are "one-click" routines for *visualizing noncovalent interactions* (DISPLAY *Contacts*), *cation-pi interactions* (DISPLAY *Cation-pi*), salt bridges (DISPLAY *Salt Br.*), and rolling-probe, solvent accessible surfaces (DISPLAY *Surface*). Evolutionarily conserved or hypervariable residues can be identified with *ConSurf*, which is integrated with PE.

ix. These and other features make PE 2 much easier to use, and much more powerful than PE 1 or RasMol. The beta-releases of PE 2 were used thousands of times per week and downloaded dozens of times per day consistently throughout 2000–2002.

x. PE 2 comprises over 60,000 lines of html, javascript, and RasMol script, making it three times the size of PE 1.
In amount of program code, it now rivals the size of RasMol. (Chime has well over 100,000 lines of C++.)

d. Chronology. Protein Explorer is built upon and depends wholly upon MDL Chime, which in turn was built upon the open-source code of RasMol.

e. A full list of novice and advanced features are available: http://www.umass.edu/microbio/chime/explorer/pe_v_ras.htm

IV. Review:
Protein explorer is one of the most widely used visualization tools for protein structures. Protein Explorer is free software for visualizing the three-dimensional structures of protein, DNA, and RNA macromolecules, and their interactions and binding of ligands, inhibitors, and drugs. Protein explorer is easy to use, free, and widely used. It can be used be novice users as well as experts. The RasMol website has been visited by over *250,000 people from 115 countries.*

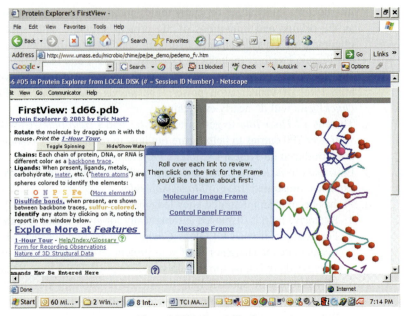

Figure 2.143.   Protein Explorer

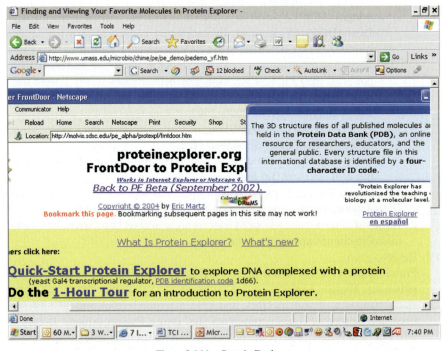

Figure 2.144.   Protein Explorer

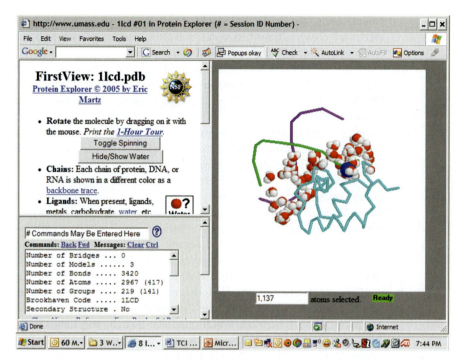

Figure 2.145.   Protein Explorer

2.32   SCHRÖDINGER, LLC

  I. Schrödinger, LLC (http://www.schrodinger.com/)
 II. Product Summaries: Schrödinger provides its FirstDiscovery software suite to
      address and aid in the discovery process in industry. First Discovery runs under
      Unix and Linux and it includes several programs: Glide, Jaguar, Liaison,
      LigPrep, MacroModel, Maestro, Mopac 2002, pKa predictor, Phase, Prime,
      QikPro, Qsite, and Strike. For the Microsoft Windows platform Schrödinger
      offers CAChe, BioMedCAChe, ChemFrontier, MaterialsExplorer, QikPro, Titan,
      and WinMOPAC.
III. Key capabilities and offerings:
      a. *FirstDiscovery*: This suite includes tools for the structure-based design of
         combinatorial compound libraries and their high-throughput analysis (dock-
         ing, ADME, etc.).
           i. *Glide*. This program allows for the fast and the automated high-throughput
              docking of compound ligand libraries to a receptor's binding site. Glide
              carries out conformational searches with Monte Carlo sampling using
              GlideScore to identify the best-binding ligand conformation.
          ii. *Jaguar*. This program is focused on *ab initio* calculations capable of
              carrying out high-level theoretical calculations. Jaguar can calculate pKa

values using the B3LYP/6–31G** geometry optimization method, electronic charges, transition states, solvation energies as well as bond forming and bond forming reactions and energies.

iii. *Liaison*. This program is used in structure-based drug design to estimate the free energy of binding between a receptor and a ligand. It uses the OPLS-AA force field to calculate energies taking into consideration solvation effects.

iv. *LigPrep*. LigPrep provides high-quality 3D molecular models from 2D or 3D molecular structures. LigPrep can add hydrogen atoms, remove counter ions, calculate and generate protonation states, minimize molecular structure, and apply filters (e.g., molecular weight, functional group) among other capabilities.

v. *MacroModel*. This program suite is used for the molecular mechanics, conformational analysis, molecular dynamics, and free energy calculations of small and medium-sized organic molecules in both gas and solution phase. MacroModel uses several force fields such as AMBER, MM2, MM3, Amber94, MMFF, MMFFs, OPLS, and OPLS-AA. For conformational analysis MacroModel uses several algorithms such as Monte Carlo Multiple Minimum, Systematic Unbounded Multiple Minimum, Pure Low-Mode Conformational Search, and Large-Scale Low-Mode Conformational Search. And for continuum solvation calculations MacroModel uses a Generalized Born/Surface Area approach.

vi. *Maestro*. This is the graphical user interface (GUI) for Schrödinger's computational platform. Maestro allows the user to execute processes via the use of a mouse (graphical interface) and also via text commands.

vii. *Mopac 2002*. Mopac is used to carry out semiempirical calculations to predict chemical properties and reactions in gas, solution or solid-state. Mopac 2002 can use several semiempirical Hamiltonian methods including MNDO, MINDO/3, AM1, PM3, MNDO-d, and PM5. Mopac 2002 comes with the MOZYME algorithms integrated for the fast calculations of electronic properties of proteins, polymers, semiconductors, and crystals.

viii. *pKa predictor*. The pKa prediction module uses *ab initio* quantum chemical methods to predict pKa of a wide range of classes of compounds in aqueous media.

ix. *Phase*. This module allows for the calculation and use of pharmacophore models for molecular database mining and QSAR elucidation. The program creates 3D databases, executes systematic explorations around rotatable bonds calculating associated conformational energies for the identification of plausible pharmacophores.

x. *Prime*. This program is used for protein structure predictions. The user can carry out low- and high-homology modeling predictions with the option to provide manual adjustments to fine-tune the results.

xi. *QikPro*. QikPro allows the fast calculations of ADME properties of drug candidates. Among the properties that QikPro calculates include polarizability (A^3), free energy of solvation (hexadecane, octanol, water), solubility, blood-brain barrier permeability, skin permeability, MDCK and Caco-2 cell permeability, and 2-D and 3-D QSAR descriptors.

xii. *Qsite*. This program uses mixed quantum mechanical/molecular mechanical methods to calculate protein-ligand interactions at an active site. Qsite accounts transition metals contained in biologically active proteins (i.e. cytochrome P450, matrix metalloproteases, Zn finger proteins) for ligand-enzyme energy interactions. For quantum mechanical calculations QSite can use different theory levels such as HF, DFT, and LMP2. For molecular mechanics calculations Qsite uses the OPLS-AA force field, and for solvation effects Qsite takes the continuum solvation model using the Poisson-Boltzman Solver.

xiii. *Strike*. This software is for statistical modeling and QSAR. It can import molecular properties and descriptors, perform multiple regression analyses (e.g., partial least squares, principal component analysis, multiple linear regression) to derive QSARs.

f. For the Microsoft Windows platform Schrödinger offers the following programs:

i. *CAChe*. CAChe is a computer-aided chemistry modeling package for life sciences, materials and chemicals applications. CAChe provides all the capabilities found in MOPAC 2002 handling a broad range of transition metals and molecules of up to 20,000 atoms including proteins and polymers. CAChe also comes with the Protein Sequence Editor for the alignment of two or more proteins as well as the construction of new protein structures using an amino acid structure repository.

ii. *BioMedCAChe*. This protein-to-drug discovery modeling package is used to devise structure-activity relationships, optimize compound leads to improve both activity and bioavailability. BioMedCAChe can process compound libraries for docking experiments, carry out calculations of descriptors to determine ADME and QSAR/QSPR predictions, and run "rule-of-five" calculations/filters.

iii. *ChemFrontier*. This program is for the creation of publication-quality 2D and 3D chemical structures, reaction diagrams and the interconversion of 2D–3D molecular structures. ChemFrontier can interact with Microsoft Office 97 via cut and paste.

iv. *MaterialsExplorer*. This molecular dynamics package has been particularly designed for materials science research. MaterialsExplorer can take into account temperature, pressure and rigid-body factors for the calculations.

v. *QikPro*. QikPro offers the same capabilities offered in the Unix/Linux version (see above).

vi. *Titan*. This is the graphical interface (GUI) to build and manipulate molecules, display results, and carry out molecular mechanics minimization, semi-empirical and high-level theory calculations for compound studies and QSAR analysis.

vii. *WinMOPAC*. WinMOPAC, based on MOPAC, allows the modeling, visualization, and calculation of molecular properties.

IV. Review:

Schrödinger's computational packages cover a broad range of research applications: theoretical chemistry, medicinal chemistry, combinatorial chemistry, structure-based drug design and protein design, and materials science. The available packages are well integrated under a common interface regardless of the operating system (Unix, Linux, and Windows). The integration of MOZYME provides the means to carry out fast semi-empirical calculations on large molecules that otherwise would be computationally prohibited to attempt. The availability to use experimentally determined drug ADME/ blood-brain barrier permeability/ MDCK and Caco-2 cell permeability properties found in QikPro allows for the filtering and prediction of synthesized and virtual compound libraries to identify promising lead compounds for their synthesis and/or further optimization. Glide provides an excellent, fast, and validated engine for ligand-receptor docking studies. It would very helpful to have information if Schrödinger's computational packages that are available for Linux are also functional under the new Linux-based Mac OS. It would also be of great help for the scientist if it were possible to add customized in-house obtained ADME data as modules into QikPro. QikPro would also benefit if other data such as efflux pump data (i.e., P-Glycoproteins) were implemented to account for additional relevant mechanisms that affect drug delivery and availability. Schrödinger at this point in time does not include or offer modules to carry out library enumeration or pharmacophore analysis and prediction.

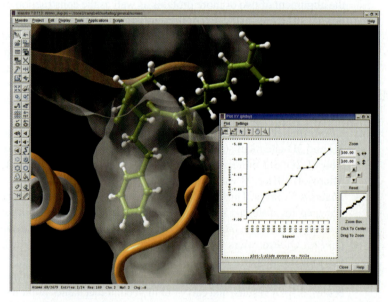

Figure 2.146.   The Maestro interface includes tools for structure visualization and data analsysis. Here, Maestro illustrates a ligand (green) docked to stromelysin (orange ribbons and white surafce). The docking scores of a series of stromelysin inhibitors have been plotted within Maestro's project facility.

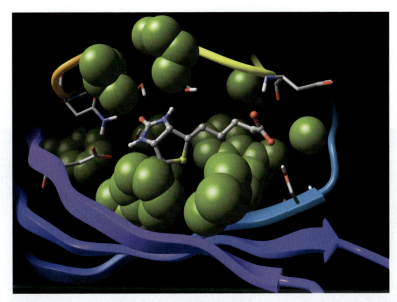

Figure 2.147. Glide's scoring function, which accounts for hydrophobic enclosure and accurately rewards hydrogen bonding interactions, is able to uniquely explain the high affinity with which biotin binds to streptavidin. Here, enclosing hydrophobic groups are colored green, while H-bonding protein residues are renered as tubes

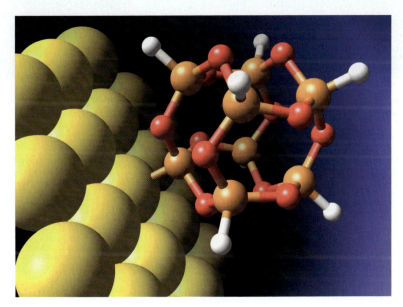

Figure 2.148. Jaguar has been successfully applied to a wide range of research projects. In the example shown here, Jaguar's unique initial guess for transition metals allowed it to rapidly predict vibrational frequencies for siloxane clusters on gold surfaces

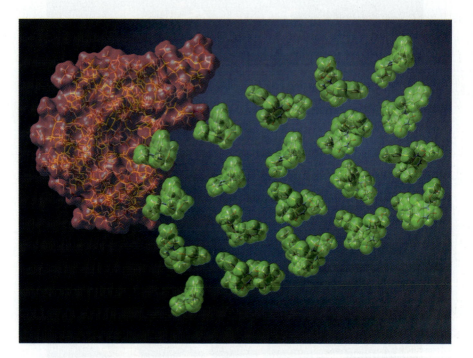

Figure 2.149.   (Liaison). Given a protein structure and series of ligands with known activities,
Liaison uses the Linear Interaction (LIA) method to generate a model for predicting binding
energies. In the example shown here, Liaison predicts the activities of 20 HIV-RT
inhibitors with an accuracy of about 1 kcal/mol

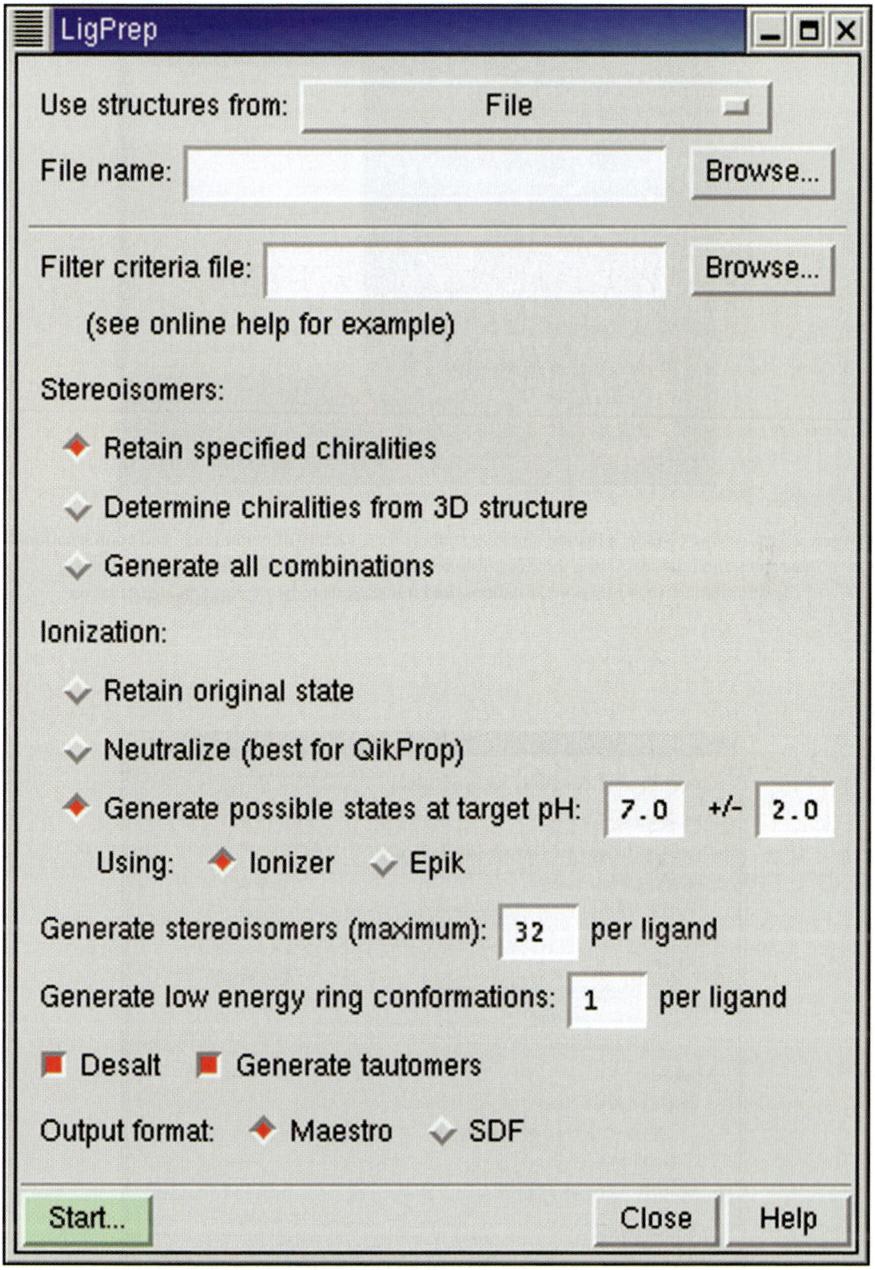

Figure 2.150.   Ligprep is used for the rapid conversion of 2D structure files into accurate
3D models suitable for use in database screens and other applications. At the user's discretion
Ligprep can generate multiple protonation states, enumerate chiralities, and generate
tautomeric forms of the ligand

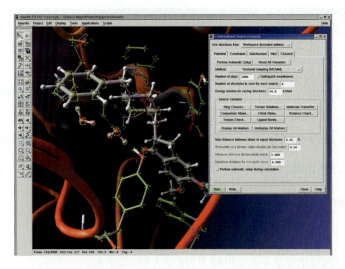

Figure 2.151. MacroModel's features include support for a variety of commonly used conformational search methodologies. Using the MacroModel interface integrated within Maestro, the user can automaticly select torsions, rotations, and translations to be explored in a calculation

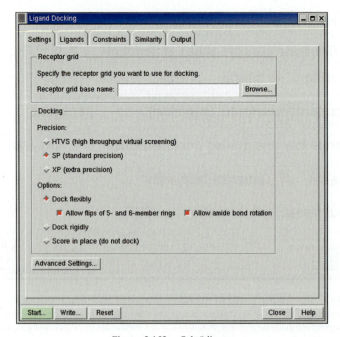

Figure 2.152.   Schrödinger

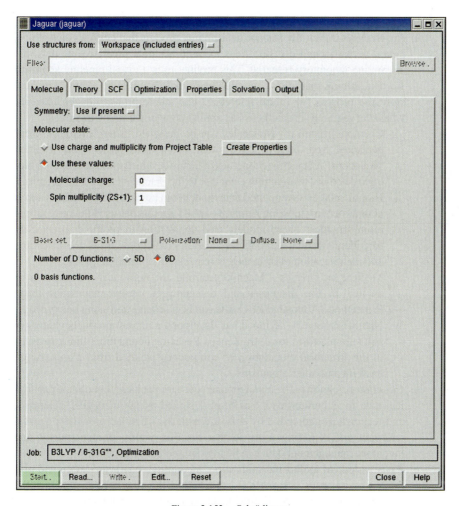

Figure 2.153.   Schrödinger

2.33   SCINOVA TECHNOLOGIES

I. Scinova Technologies http://www.scinovaindia.com/
II. Product Summaries: Scinova Technologies designs software products to help drug discovery researchers create enriched sets of probable drug candidates from large corporate chemical databanks. This objective is achieved by developing and deploying customized data warehousing and data mining solutions for biotech, drug discovery, diagnostics and chemical companies. Custom software can be created for specific customer needs and they undertake simulation and modeling studies for ongoing, dynamic biological and chemical data /processes. Main product offering consists of *Prometheus*, *Chemlab* and *Rx*.

III. Key capability and offering:
  a. *Prometheus* is their life science data-mining product that employs algorithms, neural networks and decision trees to build accurate predictive models that streamline laboratory research. This product uses a visual pipeline approach for data mining that allows researchers to exit the pipeline at any stage and store the process for future reference. The whole project can be reloaded whenever required. Prometheus has applications in the following areas:
      i. Chemoinformatics – Molecules can be designed using "chemical rules" generated from known activity data. Prometheus presents the resultant "equations" relating the properties of newly designed molecules to their biological activity in an intuitive manner to the practicing medicinal chemist.
      ii. Bioinformatics – Annotated data models can be built in Prometheus and is complimentary to string similarity-based approaches used in conventional bioinformatics. These models find application in promoter prediction, fold identification, sub-cellular location prediction, GPCR classification, and identification of drug targets (e.g. essential bacterial genes).
      iii. Medical Informatics – Machine learning and non-linear methods can be applied to the diagnosis and detection of cardiac anomalies using Prometheus. Annotated ECG data can be also analyzed using this package.
      iv. Natural Products – Scinova has developed a natural products chemistry software platform for Astra Zeneca Research Foundation, India, to assist in the structural elucidation of compounds isolated from plants, fungi, bacteria and other organisms.
  b. *Chemlab* is a knowledge management software package that collates analytical data for laboratories. It has been designed by lab managers, scientists, and technicians and coded by software experts. Chemlab provides powerful querying and reporting tools so that reaction yields, product purity, and inventory management data can be rapidly and easily generated.
  c. *Rx* is a medical toolkit for calibration, annotation, characterization and prediction based on ECG signals. It can handle a large number of signals at the same time and can be configured to sit on a database of signals.
      i. Annotation
        Rx can take in data in a variety of standard formats including image formats. It can annotate various parts of the signal, such as qrs intervals, T wave, left bundle and right bundle block beats to provide a more transparent view of the data.
      ii. Calculations and characterization
        After annotation, *Rx* calculates various characterizers for the signals. These characterizers fall in three broad categories.
        1. Physiological: these include heart rate, stroke volume, blood pressure, Rr intervals, vector cardiograms, and 3-D vector cardiograms.
        2. Statistical Means: variances, peak to peak amplitudes, Fourier transforms, correlations and cross correlations, histograms, and medians. It can also apply low pass/high pass filters to the signal to sample and vary sampling frequencies.

    3. Linear: a number of entropies, dimensions, Renyi measures, multi-fractal and fractal measures, and surrogate based quantities.

iii. Prediction

These features are then passed on to machine learning algorithms to classify the signals. In particular, multifractal properties are a clear indicator of arrhythmia in ECG signals. The onset of deformity in the ECG signal has a signature pattern that can be used to predict the onset of a heart attack. Scinova proposes to exploit this feature of *Rx* as an early warning system for intensive care units.

IV. Review:

The benefits of presenting chemoinformatics data in an intuitive manner make Prometheus an obvious asset to researchers. The customizable software being developed for chemistry of natural products should enable scientists to rapidly process large spectroscopic data sets and solve the structure of new isolates more efficiently. The machine learning aspect of the medical informatics component holds significant future promise. Additionally, the characterization aspect of *Rx*'s toolkit is particularly attractive for medical practitioners. Fully developing the early warning system component of *Rx*'s toolkit would increase its usefulness in ICU's and the medical community.

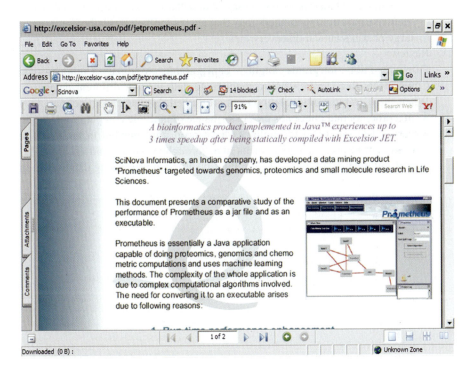

Figure 2.154.   Scinovia

2.34   SCITEGIC

  I. SciTegic, http://www.scitegic.com

 II. Product Summaries:

    a. Pipeline Pilot (http://www.scitegic.com/products_services/pipeline_pilot.htm) is a high-throughput data analysis and mining system that allows one to define and perform a complex series of operations on extremely large data sets in real time. Automated *data analysis pipelines* can be saved for future re-use or shared with a broad community of users within the enterprise across the web.

III. Key capabilities and offerings:

    a. Ability to process entire data sets so analysis is not limited to the tables of a conventional database system.

    b. *Statistical methods* can be employed which consider numeric, text, categorical, binary and fingerprint data simultaneously.

    c. Ability to pipeline together several disparate data sources, or even query on derived or calculated information.

    d. Ability to evaluate or process multiple data sets as if they were a single source.

    e. Ability to run ad-hoc queries using derived information calculated on the fly.

    f. Ability to query raw flat-file or web data as if it were already stored in a fully annotated database.

    g. Scitegic distributes NovoDynamics' ArborPharm(TM) product that uses a recursive partitioning methodology that allows for high throughput mining of large pharmaceutical discovery data sets.

IV. Review:

Pipeline Pilot integrates highly configurable Modular Computational Components that can be used to manipulate data in/from different databases. When the Computational Components are linked together into pipelines, they form a protocol that integrates the performance of multiple computational steps. Powerful data modeling technologies have been implemented in the Pipeline Pilot system to take advantage of its high-throughput nature and accommodation of multiple data types. Pipeline Pilot is aimed at processing data based on user-defined rules or filters, but for chemical reaction transforms where the compound enumeration can follow complex rules, parallel reactions, and crucial compound definitions such as stereochemistry, salt forms, tautomers, etc., the use of Pipeline Pilot is non-trivial. Pipeline could be used for alternative applications if it is complemented with additional tools and environments. The technology is partially open for such user-defined extensions.

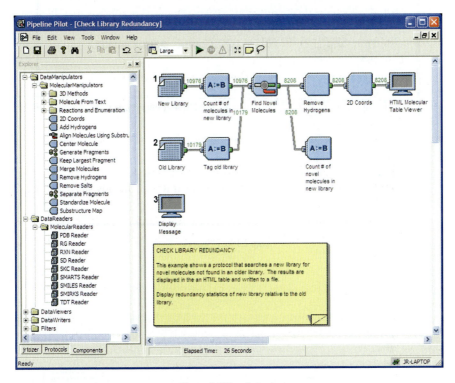

Figure 2.155.   Scitegic

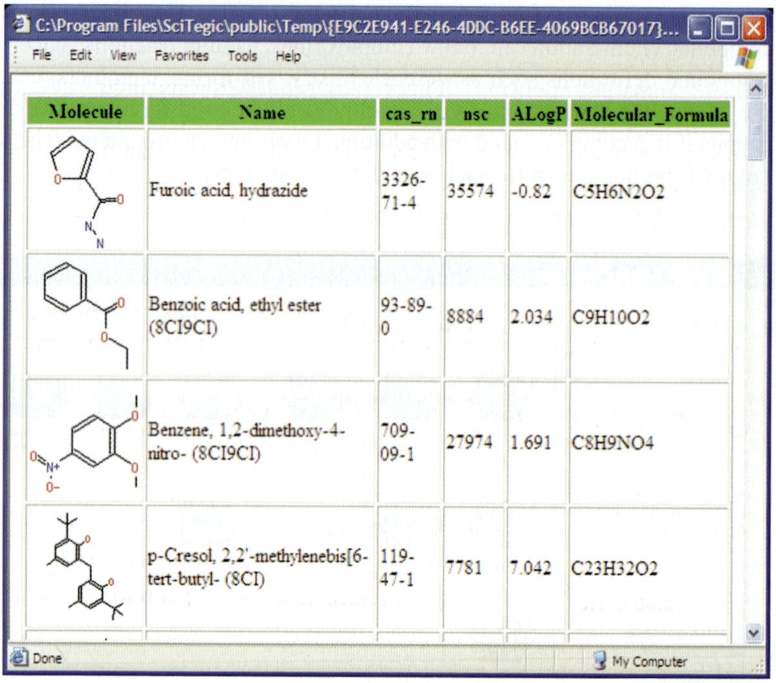

Molecule	Name	cas_rn	nsc	ALogP	Molecular_Formula
	Furoic acid, hydrazide	3326-71-4	35574	-0.82	C5H6N2O2
	Benzoic acid, ethyl ester (8CI9CI)	93-89-0	8884	2.034	C9H10O2
	Benzene, 1,2-dimethoxy-4-nitro- (8CI9CI)	709-09-1	27974	1.691	C8H9NO4
	p-Cresol, 2,2'-methylenebis[6-tert-butyl- (8CI)	119-47-1	7781	7.042	C23H32O2

Figure 2.156. Scitegic

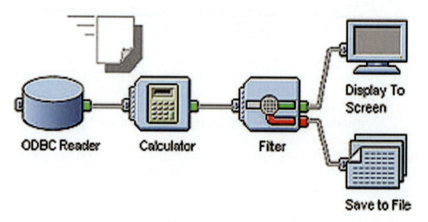

Figure 2.157. Scitegic

2.35 SIMULATION PLUS, INC.

I. Simulations Plus, Inc. (http://www.simulations-plus.com/)

II. Product Summaries: Simulations Plus develops simulation and predictive modeling software for *in silico* compound screening and for preclinical and clinical drug development in the area of Absorption, Distribution, Metabolism, Excretion and Toxicity (ADMET). The available applications include: GastroPlus, ADMET Predictor, ADMET Modeler, DDDPlus, and MembranePlus.

III. Key capabilities and offerings:

a. GastroPlus: This package is a platform that integrates *in silico, in vitro*, and *in vivo* data. It simulates gastrointestinal absorption, pharmacokinetics, and pharmacodynamics for intravenous and orally dosed drugs in animals and human. This suite includes:

  i. Physiologically-based Advanced Compartmental Absorption and Transit (ACAT) models for several species: human (fasted and fed), beagle dog (fasted and fed), rat, mouse, rabbit and cat.

  ii. Intravenous bolus and infusion, immediate release tablet and capsule, and controlled release (dispersed, integral tablet, and gastro-retentive) dosage models for simulation of plasma concentration vs. time profiles.

  iii. Database fields for all relevant physico-chemical (log P, pKa, solubility), and biopharmaceutical properties (permeability, transporter and enzyme gut distributions and Michaelis-Menton parameters).

  iv. Deconvolution and correlation of *in vitro* dissolution and *in vivo* release data.

  v. Calculation of the pH-dependence of partition coefficient (log D), solubility, and dissolution rate.

  vi. Available add-on modules include:

    1. *PDPlus* module to fit standard pharmacodynamic (PD) public domain models (both direct and indirect) to user supplied PD data.

    2. *Metabolism and Transporter* module that uses known gut distribution of enzymes and transporters (both influx and efflux) in combination with V_{max} and K_m parameters to predict gut and liver metabolism, and the nonlinear dose dependence for substrates of influx, and efflux transporters.

    3. *The Optimization Module* aids in the optimization (fitting) of a wide variety of model parameters including physiological, pharmacokinetic, pharmacodynamic, and formulation variables.

    4. *PKPlus* module to rapidly fit PK parameters to IV plasma concentration-time (Cp-time) data for noncompartmental and 1-, 2-, & 3-compartment models.

    5. *IVIVC (In-Vitro, In-Vivo Correlation)* module to correlate relationships between the *in vitro* dissolution of tablets and capsules and the *in vivo* dissolution in human and animal systems.

    6. *PBPK Module* allows the user to create rat (adult only) and human organ physiology (any age between 1 and 85 yo) for use in

Physiologically-Based Pharmacokinetics. Tissue-plasma partition coefficients may be calculated automatically by using only log P and fraction unbound in plasma as inputs. This module allows the user to simulate Cp vs. time profiles for drugs in discovery using only *in silico* or *in vitro* (metabolism) data inputs.

b. ADMET Modeler: This package makes it very fast and easy to generate artificial neural network ensemble and support vector machine ensemble models using molecular descriptors from ADMET Predictor for structure-to-property predictions from sets of experimental data. All models in ADMET Predictor have been generated with ADMET Modeler.

c. ADMET Predictor: This suite estimates and correlates structure-derived ADMET properties using 49 predictive models and 275 diverse descriptors for *in silico* combinatorial library screening. This suite:
   i. Accepts 2D or 3D molecular records in several formats:
      1. SMILES strings.
      2. MDL's *.sdf files.
      3. MDL's *.rdf files.
      4. MDL's *.mol files.
      5. CambridgeSoft's ChemDraw's *.mol files.
      6. Tripos Inc.'s SYBYL's MAC format.
   ii. Calculates multiprotic ionization constants (pK_a), log P, water solubilities, effective and apparent permeability, blood-brain barrier penetration, simulated fraction absorbed in human, 7 types of toxicity, plasma protein binding, volume of distribution in human, and diffusivity.
   iii. Designed to run interactively or in batch mode from either a graphical user interface, or command line.
   iv. Interfaces with SciTegic's Pipeline Pilot software for automated drug candidate screening.
   v. Other features include:
      1. Property/descriptor histograms and correlation graphing.
      2. *4D Data Mining* module utilizing the novel concept of Chemically Relevant Principal Components to visualize and analyze chemical space of results from ADMET Predictor.

d. DDDPlus: DDDPlus (Dose Disintegration and Dissolution Plus) is a program that simulates the *in vitro* dissolution of active pharmaceutical ingredients (API) and formulation excipients dosed as powders or tablets under various USP experimental conditions. This software accounts for the effects of:
   i. Physicochemical properties of the formulation ingredients under study: lipophilicity, pKa's, solubility, diffusion coefficient, and density.
   ii. Processing variables for immediate release dosage forms.
   iii. Particle size distributions for each of the formulation ingredients.
   iv. Different flow patterns and fluid velocities for each experimental apparatus.
   v. Interactions between the active ingredient and formulation excipients.

    vi. Microclimate pH-dependence of solubility and dissolution/precipitation.

    vii. Micelle-facilitated dissolution through the incorporation of surfactants in the media.

e. *MembranePlus (under development)*: MembranePlus is a detailed mechanistic simulation of experiments used for artificial membrane permeability (PAMPA) and epithelial cell culture permeability (Caco-2, MDCK) assays. MembranePlus allows the user to simulate the kinetics mass transfer due to molecular diffusion in the extracellular compartment, apical and basolateral membrane compartments, and the intracellular compartment of epithelial cells in culture. The software accounts for the effects of:

    i. Partition coefficient (log P), pKa, and pH, on entry rate into the apical and basolateral membranes, and accumulation of drug in each bilayer

    ii. Unstirred water layers on both apical and basolateral sides of the membrane.

    iii. Molecular concentration on mass flux across the epithelial cell monolayer or the artificial membranes found in PAMPA assays.

    iv. Paracellular permeability in cell culture models.

    v. Transcellular and paracellular contributions to human effective permeability for all regions of the small and large intestines.

    vi. Carrier-mediated influx and efflux of molecules in the context of a virtual *in vitro* permeability assay.

    vii. Intramembrane concentration kinetics on membrane retention and rate of approach to steady-state.

    viii. User-sampling from the donor or acceptor wells.

The results of the simulations provide the user with simulated values for cell culture or PAMPA permeability and with estimates for human permeability in all areas of the human small and large intestine as well as allowing for plotting of the kinetics in all compartments of the simulation.

IV. Review:

Simulations Plus' products provide to the medicinal chemist and combinatorial chemist a comprehensive set of tools to generate and analyze in-house generated chemical structure-molecular properties-biological profiles and compare it with public-domain ADMET derived models to map relevant chemical space, filter out small and large compound sets to select more promising candidates for lead generation and/or lead optimization. With all these packages available for the Windows 95/98/NT/2000 & XP platforms a great percentage of the scientific community would likely meet the minimum system requirements to use the above packages. Batch-mode capabilities are incorporated into almost all of these products to allow the user work on multiple projects in unattended mode. It would be a great addition to have the above packages available to the Mac OS platform which is widely used particularly in academia.

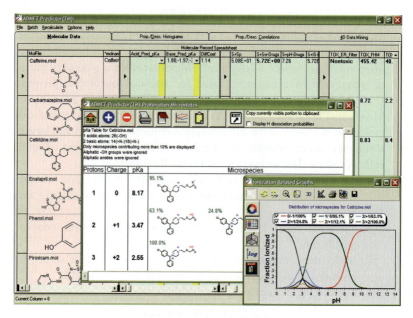

Figure 2.158.   Simulations Plus

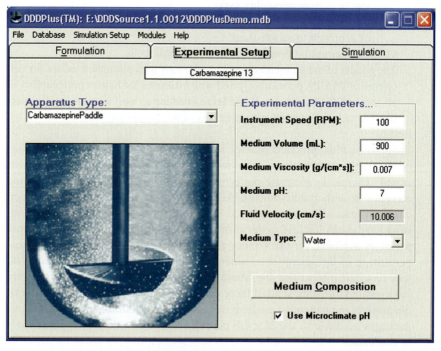

Figure 2.159.   Simulations Plus

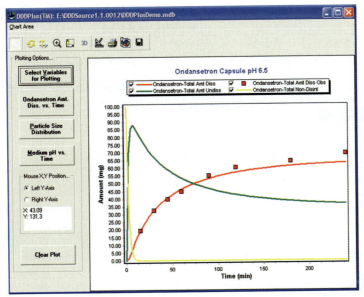

Figure 2.160.   Simulations Plus

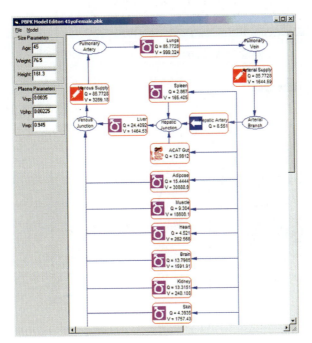

Figure 2.161.   Simulations Plus

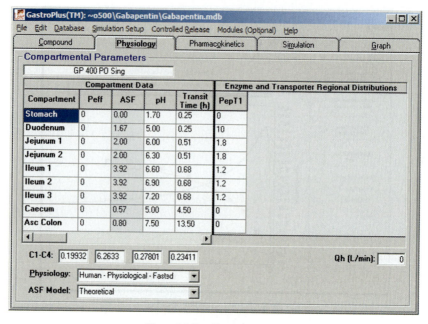

Figure 2.162.   Simulations Plus

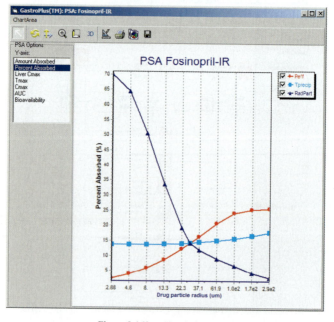

Figure 2.163.   Simulations Plus

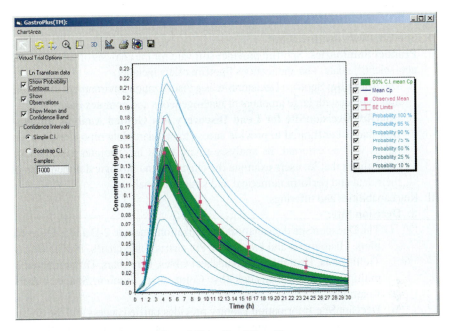

Figure 2.164.   Simulations Plus

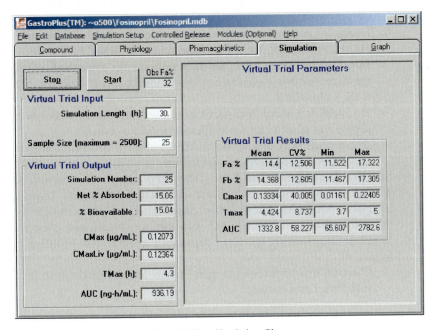

Figure 2.165.   Simulations Plus

2.36   SPOTFIRE

I. Spotfire (www.spotfire.com)
II. **Product Summaries:** Spotfire provides software for the analysis of scientific and non-scientific data. The applications Spotfire offers include:
   a. **Decision Site:** Spotfire DecisionSite is a visualization environment to explore and interact with large amounts of heterogeneous and complex data.
   b. **Spotfire DecisionSite for Lead Discovery is a Guided Analytic application** that is configured to provide access to standard data repositories, internal as well as external. Its analysis environment incorporates visualization capabilities that let users examine the chemical and biological dimensions of their data and perform numerical and structural analyses.
III. Key capabilities and offerings:
   a. **Decision Site:**
      i. The DecisionSite data analysis visualizations include 2-D and 3-D scatter plots, Histograms and bar charts, Pie charts, Line charts, Profile charts, Trellis plots, Heat maps, Spreadsheet tables, Error bars, Differentiated in multiple visible dimensions (by: Color, Size, Symbol, Shape, Labels, Rotation).
      ii. DecisionSite Information Library provides unified access and the ability to visually integrate data, in local or remote file systems, in relational databases and on Web sites. Data is accessible from a range of sources such as Microsoft Excel, text files (comma and tab delimited), XML files, and industry-standard databases (e.g., Oracle, Microsoft SQL Server, MySQL).
      iii. Decision Site can format or export the data in several formats such as HTML, Powerpoint, Word, Excel and comma separated files.
      iv. DecisionSite Computation Services supports the integration of algorithms and models, including commonly used statistical packages such as SAS, S-plus and R.
      v. DecisionSite Guided Analytics can capture an analysis workflow and deploy it as guided analysis application for use by others
      vi. DecisionSite Developer allows the development of customized analytical applications based on Spotfire's platform.
   b. **Spotfire DecisionSite for Lead Discovery** allows the searching and profiling of compounds to elucidate and create SAR reports. Users can integrate SAR tables into an active visualization environment with tools to perform hierarchical and ISIS structure key-based data clustering. Spotfire DecisionSite for Lead Discovery enables the handling and mining of:
      i. Chemical structures retrieved from Elsevier MDL's ISIS database and the ISIS Direct data cartridge.

ii. Chemical properties and information retrieved from CAS SciFinder.

iii. HTS data stored in IDBS's ActivityBase system.

IV. Review:

Spotfire provides a data analysis and visualization tool for essentially any data, including chemical data. Because Spotfire is so generally applicable, many domain-specific applications either need to be custom made with a DecisionSite Developer license or obtained through combining Spotfire with other commercially available products. Companies that provide their products compatible with Spotfire include Elsevier MDL, CAS, CambridgeSoft, and Daylight, among others. For the analysis, clustering, and 2D and 3D visualization of large data sets Spotfire provides a solid and robust platform with the opportunity to develop customized applications when needed.

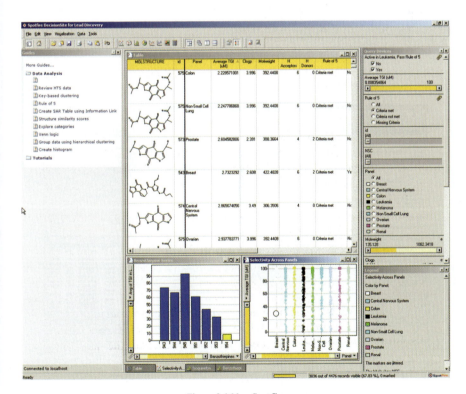

Figure 2.166.   Spotfire

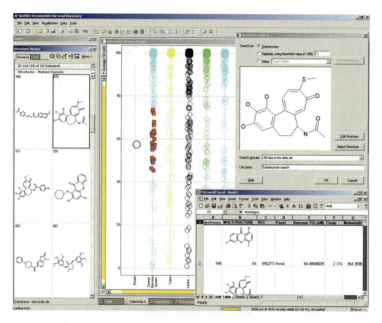

Figure 2.167.   Spotfire

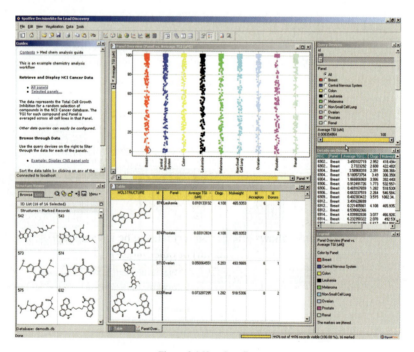

Figure 2.168.   Spotfire

2.37.   SUMMIT PK

I. Summit Research Services (Summit PK); http://www.summitpk.com/

II. Product Summary

    a. **PK Solutions** is an automated Excel-based program to compute single and multiple dose pharmacokinetic data analysis of concentration-time data from biological samples (blood, serum, plasma, lymph, etc.) following intravenous or extravascular routes of administration. The program provides tables of the most widely used and published pharmacokinetic parameters. PK Solutions calculates results using noncompartmental (area) and compartmental (exponential terms) methods without presuming any specific compartmental model. Multiple dose and steady state parameters are automatically projected from single dose results.

    b. **Metabase** is a laboratory data management software system designed specifically to meet the needs of laboratories conducting ADME, pharmacokinetics, and toxicokinetics studies. Starting with the generation of a sample list for an entire study, Metabase serves as the basic tool to acquire, organize, manage, calculate, store, and report study results. Key features of the program are its organization (which follow standard laboratory workflow procedures), its data handling, and its integration with other productivity software. The program is based on Microsoft Excel and can be custom tailored to fit each client's work place or modified to meet changing needs and requirements. **Metabase** comes as a turn-key system for radio-analytical sample processing and liquid scintillation counting data, but it can be custom modified for other analytical methods. Metabase is designed to meet GLP guidelines and comply with FDA, EPA, EEC software development and validation requirements. The system is provided with complete system life-cycle documentation. Test files are supplied for use in validation.

III. Key capabilities and offerings:

    a. PK Solutions:

        i. Works with both concentration-time data and imported exponential terms.

        ii. Prediction of more than 75 single and multiple dose pharmacokinetic results using intravenous or extravascular dose data. PK parameter calculations are based on (1) curve-stripping to derive the exponential terms that describe the blood level curve, and (2) area under the curve calculations.

        iii. Dynamic links allows graphs and formulas to be instantly updated when changing data (e.g., dose interval) or other inputs.

        iv. Captures up to 256 data sets on a single, flexible form. Can print and export reports, data tables, and graphs in color or black and white.

    d. Metabase:

        i. Metabase can manage data from sampling individual animals, time-based composites from groups of animals, multiple dose levels, and variable specific activities.

ii. Metabase can import data from files obtained from a liquid scintillation counter or by direct connection with the equipment.

iii. Metabase can automatically prepare a Sample Processing spreadsheet to be linked to auto-balances for direct transfer of weight measurements.

iv. Data files can handle up to 16,000 records and use multiple files for any study.

v. Radioactivity calculations to predict initial dose calculations and radioactivity required, the limit of detection and analytical sensitivity for a given specific activity, and other useful values.

vi. Pharmacokinetics – Metabase results can be passed directly to Summit's PK Solutions software for pharmacokinetics data analysis.

vii. Reports can be generated to prepare data tables with appendix and report quality formats.

viii. Metabase can be customed designed to fulfill core functions associated with metabolism sample management and data acquisition, analysis and reporting functions.

ix. Metabase complies with GLP guidelines.

x. SDLC Documentation – Metabase comes with a complete system development life cycle documentation package to make validation a breeze. The minimum validation documentation set includes Requirements Analysis, System Design Specifications, Version History, Source Code, User Manual, Test Procedures, and Test Files. All documentation is available on a CD or as printed material.

xi. Metabase files can be transported on diskettes or other portable medium (or via e-mail) to be worked on elsewhere.

IV. Review:

Summit PK software consists of Microsoft Excel programs for ADME, pharmacokinetics, and toxicokinetics studies. These programs can run on PCs using any version of Windows (Windows 3.x, 9x, NT/2000, XP) as well as on Macintosh computers. PK Solutions is designed to make pharmacokinetics data analysis easy to perform, and compiles 75 of the most commonly used pharmacokinetic parameters, including steady state and multiple dose results projected from single dose data. Summit's Excel-based programs provide a familiar working environment with extraordinary capabilities for additional analyses, importing and exporting of data, and preparation of presentation materials and reports.

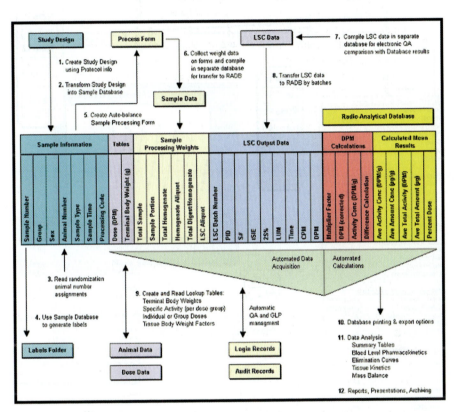

Figure 2.169. Summit PK

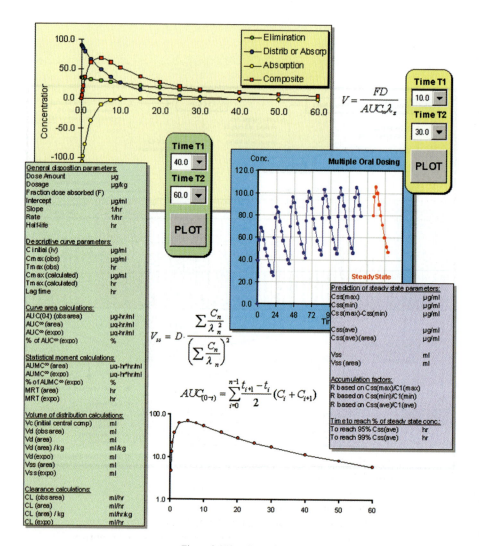

Figure 2.170.   Summit PK

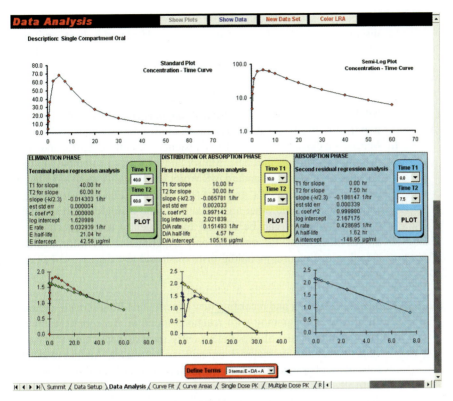

Figure 2.171.   Summit PK

2.38 SYMYX

I. Symyx Inc. (http://www.symyx.com)

II. **Product Summaries:** Symyx Software includes the following:

a. Electronic laboratory notebook software that provides uniform IP protection and project data management across the enterprise, along with tailored user interfaces for scientists in medicinal chemistry as well as pharmaceutical and chemical development.

b. Experimental design and broad data capture capability applicable to conventional and automated research.

c. Vendor-independent laboratory automation for automated research

d. Enterprise-wide real time data warehousing, querying and reporting

III. Key capabilities and offerings:

a. Symyx Software Electronic Lab Notebook (ELN) applications allow researchers to design and plan both conventional and array-based experiments in both regulated and non-regulated environments.

i. Compose automatically writes batch instructions and then uses Record to electronically capture the data. Instead of using text to describe procedures, scientists can create "structured procedures" consisting of discrete, well-defined operations such as add material, heat, and mix. These structured procedures are then automatically converted into batch instructions using standardized language. Batch data is then captured electronically using the application Record to automatically generate instructions, calculations and procedure scaling. Uniform creation of batch instructions with standardized language facilitates batch instruction quality independent of individual scientists' writing quality.

ii. Matrix is a chemically-aware ELN designed for both scientific and engineering experiments and data. Matrix allows users to assemble multiple documents, including standard MS Word, Excel, and PowerPoint files, as well as XML files, PDFs, graphics files, movies, sound files, characterizations, spectra, and other analytical data, into a single notebook entry. Custom templates can be created for frequently used procedures or unique documents for one-time-only experiments. All information is stored in one record, including instrument and test data as well as data analysis.

iii. Reaction Planner enable a scientist to explore synthetic pathways, quickly designing and recording new experiments from scratch, or, more efficiently, by beginning with reactions found in a corporate database. The repetitive tasks of experiment entry such as generating a material table and calculating material amounts are automated. Adding reactants, solvents, and products and their physical properties is facilitated by automatically searching and retrieving information from the database. Scientists can generate reports to provide either hardcopy notebook pages or formal intellectual property reports. Reaction Planner also includes enumeration tools to facilitate planning and execution of compound libraries directly within the electronic notebook.

iv. Vault is a searchable document management repository that enables drug development data to be shared, routed for approval, reused, and maintained under regulatory-compliant conditions. Vault stores development ELN records, in addition to associated information, including MS Word, MS Excel, MS PowerPoint, PDFs, graphics, movies, and sound files. Automated workflow management automatically routes these documents to the right people at the right time, streamlining the document approval process. Additionally, security settings, a full audit trail with version control, change control, and electronic signatures enable records to be stored and maintained in compliance with FDA 21 CFR Part 11. Scientists can find data stored in Vault using searching capabilities, including text-, molecule-, and reaction-based searches.

b. The automation applications within Symyx Software enable scientists to control, monitor, and manage data acquisition for automated laboratory instruments, semi-automated instruments and manual instrumentation. Symyx

Software was developed with vendor neutral architecture. The software is extensible by customers and third-party parties.

  i. Library Studio

    1. Symyx Library Studio software provides an environment for the specification of one, several, or hundreds of experiments. Chemicals, mixtures, and process parameters are defined and then combined in a recipe-style user interface.

    2. The resulting experimental designs are saved to the Renaissance Application Server (RAS), where they provide a record of the compositional makeup of each sample in the design. The designs also provide the process conditions under which the sample was created for assessing composition/property relationships.

    3. The Library Studio Developer's Kit further extends Library Studio by embedding Microsoft's Visual Basic for Applications (VBA) into the Library Studio application. With VBA, the user can import data from statistical packages, such as DOE, or export data for further manipulation, such as a bubble point calculation. Additionally, modifications to the Library Studio user interface are possible via custom add-ins, such as commands to produce automated designs or validate a design intended for a particular instrument.

c. Impressionist

  i. Symyx Impressionist software is a general laboratory automation package for creating and executing laboratory procedures. It provides an environment for creating procedures that implement liquid-handling and process-control workflows, including those that synthesize libraries created in Library Studio.

  ii. Scientists and engineers construct procedures graphically by arranging actions selected from a library into a logical hierarchy that defines the flow of execution. Over 130 predefined actions provide functionality, ranging from simple loops and math expressions to complex robot and instrument control. Most actions include configurable properties that control their runtime behavior and level of user interaction. Once created, procedures can be saved, shared, and repeatedly executed by laboratory technicians in a production environment.

  iii. In addition to actions, Impressionist includes over 120 resources that manage communications with individual hardware devices and expose their capabilities in a standardized way. This approach allows devices of the same type to be used interchangeably by procedures even if the underlying communications protocols are different. Configuring Impressionist for a specific workstation is simply a matter of selecting and activating those resources that correspond to the hardware devices physically connected to the computer. Hardware-specific settings are defined as resource properties.

iv. Other applications can use Impressionist behind the scenes via Microsoft's COM technology to execute procedures or directly control and monitor hardware resources. This allows the creation of custom programs that provide tailored user interfaces while delegating the specific details of device and process control to Impressionist.

v. The Impressionist Developer's Kit further extends Impressionist by embedding Microsoft's Visual Basic for Applications (VBA) into the Impressionist client application. With VBA, users can create and distribute their own custom actions and resources as well as modify the Impressionist user interface via custom add-ins.

d. Epoch

i. Symyx Epoch software provides an environment for creating customized protocols for doing library-based screening and data processing experiments. Scientists and engineers construct protocols graphically by arranging blocks in a flow chart that defines the flow of execution. Each block is further composed of actions, with over 300 predefined actions that provide data acquisition and processing functionality. In addition, these actions allow direct access to Impressionist resources and substrates, as well as the ability to execute Impressionist procedures. Actions include configurable input parameters that control their runtime behavior, and output parameters that allow the data generated by the action to be used elsewhere in the protocol.

ii. All data generated by a protocol during execution is stored to the Renaissance Application Server (RAS), linked both to the protocol and to the library design that generated the samples being screened. The population of data in experiments is controlled by parsers, which route action inputs and outputs to the appropriate parts of the experiment. Visual feedback is provided by results controls, which route action inputs and output to user-interface elements like spreadsheets and charts.

iii. In addition to actions, Epoch includes over 100 resources that manage communications with individual automation and data acquisition devices, and expose their capabilities in a standardized way. This approach allows devices of the same type to be used interchangeably by protocols even if the underlying communications mechanisms are different. Configuring Epoch for a specific workstation is simply a matter of selecting and activating those resources that correspond to the hardware devices physically connected to the computer. Hardware-specific settings are defined as resource properties.

iv. The Epoch Developer's Kit further extends Epoch's functionality, enabling the development of Epoch resources, actions, parsers, and results controls with Visual Basic .NET.

e. Symyx Software's data mining applications feature real-time data warehousing. Members of a team or an entire organization can utilize multiple query and browsing tools to simplify reporting or to search by document or chemical structure.

i. Spectra Studio: Symyx Spectra Studio software provides an environment for clustering and defining groups of spectra based on their intrinsic features. As applied to the Symyx Tools Preformulations Workflow, it is used to combinatorially compare XRD and Raman Spectra to identify crystalline forms of drug candidates.

ii. Polyview: Symyx PolyView software provides a flexible way to search for and retrieve data from the Renaissance Application Server (RAS) based on data stored during synthesis and screening activities. The user can view combinations of data from multi-step synthetic processes, which are cross-linked by the RAS to provide consistent and clear access to all data collected through the lifetime of a test sample. This data may be displayed in custom reports, including spreadsheet tables, 2D or 3D scatter plots, 2D or 3D bar charts, grid views, and library summaries. Reports can include numeric and textual data as well as an extended set of custom data types, including spectra, chromatograms, and images. Links are also provided to send data to external text files, Microsoft Excel and to Spotfire DecisionSite for additional visualization and analysis.

iii. Data Browser: Symyx Data Browser software provides web-based access to data stored in the Renaissance Application Server (RAS).

1. Data Browser provides access to experimental data based on an easy-to-use "drill-down" method, allowing the user to browse projects and libraries, and to produce document-style experiment and library design reports.

2. Data Browser can be used from any computer with a web browser; no other Symyx Software is required to view data and perform simple searches.

3. Symyx Data Browser can also be customized to produce additional reports, such as project or library summaries, or additional experiments with varying levels of detail.

iv. Reaction Pathfinder: Multi-step syntheses require the ability to see how intermediates can be manipulated to produce the desired compound. Symyx has built a series of data mining tools that use the empirical knowledge contained in the electronic notebook. These tools are the Reaction Relay and the Retro-Synthetic Tree. The Reaction Relay provides a dynamically-generated view of the entire multi-step synthesis, displaying all reactants used to make the chosen molecule as well as all the products generated from that substance. The Retro-Synthetic Tree is a dynamically-generated view of all the known synthetic pathways to make a given substance. This view displays all the reactions, and the reactants they produce, leading to the product of interest, with yield information displayed for each reaction. Both of these tools provide links to the detailed experimental procedures for each reaction.

   f. Previously components were sold by Intellichem and Synthematix including:

       i. ARTHUR Suite. This package is oriented towards the creation, organization and design of experimental procedures as well as the data generated. The applications offered with this suite are: ARTHUR KnowledgeBase and ARTHUR Reaction Planner.

      ii. StructureSearch. This package is for the substructure search of chemical structures.

IV. Review:

Through internal development and acquisition Symyx has a range of tools for informatics within the pharmaceutical and chemical industries for both classical and high-throughput experimentation. Symyx Software covers enterprise-wide issues. This includes software that serves as the core interface for individual scientists to organize their research and manage their experiments. A common back-end repository for data from conventional and automated research, our software combines and organizes information from multiple scientists spanning project teams, departments, and geographic sites. The overall goal is to assist with knowledge-driven decision making.

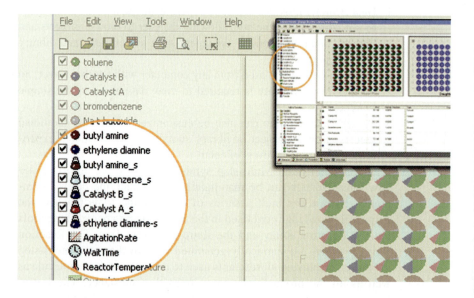

Figure 2.172. Symyx

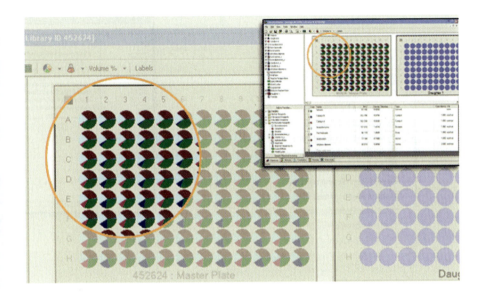

Figure 2.173.   Symyx

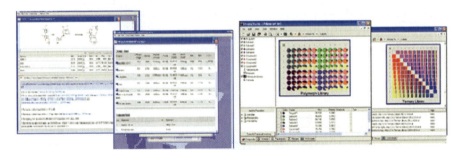

Figure 2.174.   Symyx

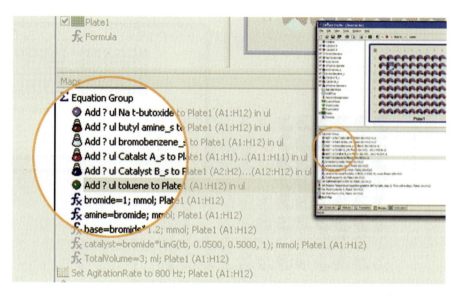

Figure 2.175. Symyx

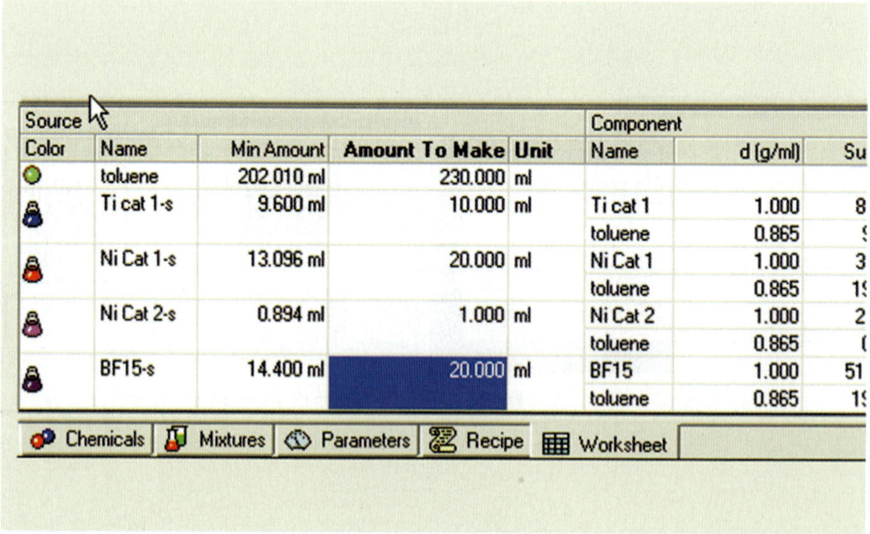

Source					Component		
Color	Name	Min Amount	Amount To Make	Unit	Name	d (g/ml)	Su
●	toluene	202.010 ml	230.000	ml			
🍶	Ti cat 1-s	9.600 ml	10.000	ml	Ti cat 1	1.000	8
					toluene	0.865	9
🍶	Ni Cat 1-s	13.096 ml	20.000	ml	Ni Cat 1	1.000	3
					toluene	0.865	19
🍶	Ni Cat 2-s	0.894 ml	1.000	ml	Ni Cat 2	1.000	2
					toluene	0.865	0
🍶	BF15-s	14.400 ml	20.000	ml	BF15	1.000	51
					toluene	0.865	19

🔴 Chemicals 🧪 Mixtures ⚙ Parameters 🎼 Recipe 🏢 Worksheet

Figure 2.176. Symyx

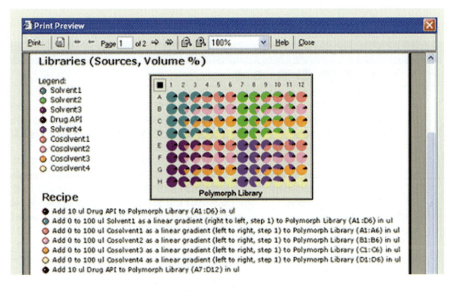

Figure 2.177.   Symyx

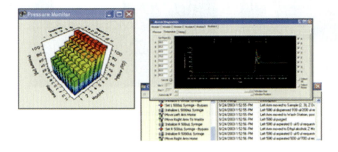

Figure 2.178.   Symyx

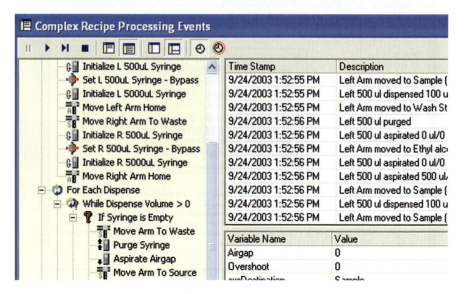

Figure 2.179. Symyx

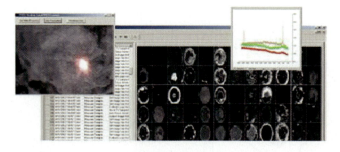

Figure 2.180. Symyx

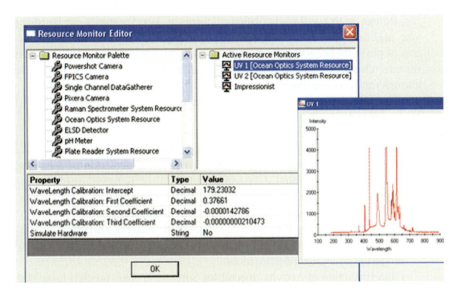

Figure 2.181.   Symyx

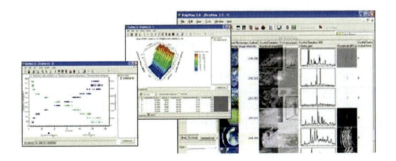

Figure 2.182.   Symyx

2.39   TIMTEC

   I. TimTec LLC (http://software.timtec.net)
  II. Product Summary: Chemical Database Management Software, Diversity, Physico-
      Chemical Properties TimTec's database management system is designed for enter-
      ing, storage, search and manipulation of chemical data. Databases can be of any
      size. Additional features include spectra processing, diversity analysis and predic-
      tion of physico-chemical properties (Log P, Log D, solubility, absorption). Users
      include chemists, biochemists, ecologists, molecular biologists, medicinal
      chemists and pharmacologists.
      a. *ChemDBsoft* Chemical database management software.
      b. *Structure Editor* included with CHED.
      c. *SLIPPER* pK Prediction Log P, Log D, Log Sw, FA.
      d. *DISCON* Dissociation Constants. pK Prediction.
      e. *HYBOT-PLUS* Hydrogen bond thermodynamics. Calculation of local and
         molecular physicochemical descriptors.
      f. *MOLDIVS* Molecular diversity and similarity estimation.
      g. *FLAME* Database on combustion of explosives and propellants.
      h. *HAZARD* Sensitivity and properties of energetic materials database.
      i. *ASTD* Reference on thermodynamic, thermochemical and thermophysical
         properties of individual substances.
      j. *PASS* Prediction of Activity Spectra for Substances. Pass predicts more than
         700 pharmacological effects, mechanisms of action, carcinogenicity, muta-
         genicity, teratogenicity and embryotoxicity on the basis of structural formula
         of the compound.
      k. *REAL* Computer modeling of complex chemical equilibrium at high pressure
         and temperature
 III. Key Capabilities and Offerings
      a. ChemDBsoft Chemical database management software can handle up 10 mil-
         lion records, support several data formats (e.g., SDF, MOL, JCAMP), stores
         molecular spectra (NMR, MS, IR), and offers tools for diversity analysis as
         well as synthesis planning.
      b. Structure Editor is included with ChemDBsoft program for the creation and
         editing of chemical structures.
      c. Slipper Prediction is a tool to precidt lipophilicity, solubility, pH-dependent
         profiles and drug absorption. SLIPPER-2001 (Solubility, LIPophilicity,
         PERmeability) can predict octanol/water partition coefficient values (log
         D) and aqueous solubility values (log Sw) at any pH ranges with the capa-
         bility of exporting the results in several formats (e.g., doc, sdf , xls, txt).
      d. DISCON comes with a database with over 16,000 pK data points to allow the
         estimatimation of pK values for ionizable compounds at all pHs. This pro-
         gram can also work in batch mode.
      e. HYBOT (HYdrogen BOnd Thermodynamics) comes with three major com-
         ponents: (i) a database of over 13,500 thermodynamic measurements of

hydrogen bonding systems, (ii) a H-bond factor database (about 60,000 entries), and (iii) a computer program that estimates structure-based H-bond factor values. HYBOT allows to import and export compounds using standard sd files.

f. MOLDIVS (MOLecular DIVersity and Similarity) is a program package to run molecular similarity and diversity calculations on large sets of compounds. This program allows:
   i. Calculation of similarity indexes.
   ii. Estimation of whole diversity of any database of chemical compounds.
   iii. Selection of diverse subset of compounds from any database of chemical compounds.

g. FLAME: is a database on combustion of explosives and propellants. The database contains experimental data on combustion of more than 700 explosives and more than 300 propellants.

h. HAZARD. This database, which is a good complement to FLAME, provides information on sensitivity characteristics of individual explosives, explosive mixtures, propellants, and pyrotechnic compositions.

i. ASTD. This database covers thermodynamic, thermochemical and thermophysical property data of about 2,500 gases and condensed species substances.

j. PASS (Prediction of Activity Spectra for Substances) uses about 30,900 biologically active compounds in the training set and around 550 names of biological activities to predict biological activity spectra including mechanisms of action, carcinogenicity, mutagenicity, teratogenicity and embryotoxicity.

k. REAL builds computer modes of complex chemical equilibrium at high pressure (up to 600–800 MPa) and temperature (up to 6,000 K).

IV. Review:

Although TimTech focuses primarily on providing compounds for screening, they offer ther CHED software package for life sciences. Because this software was primarily developed to be used for ordering TimTec's compounds certain capabilities may need to be integrated with internal organization-specific software. However for searching the TimTec database the freely available tools are easy to use and rapidly return hits from structure searches. TimTec's CHED software is mainly available for the Windows operating system, although a few components are available for UNIX. Because of this Mac or Linux users will likely have to either look for other alternatives or attempt to use CHED under a Windows emulator.

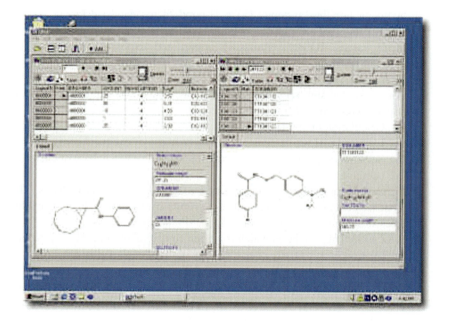

Figure 2.183.   TimTec

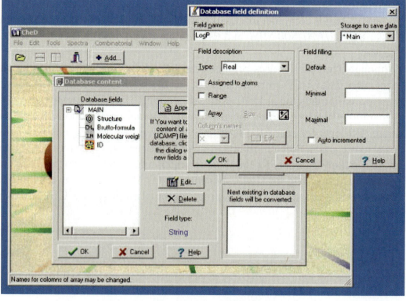

Figure 2.184.   TimTec

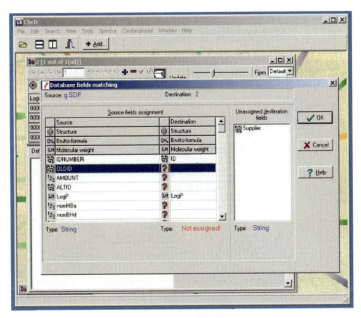

Figure 2.185.   TimTec

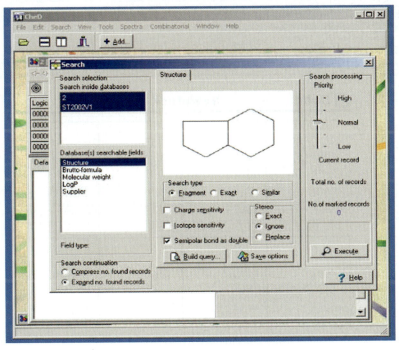

Figure 2.186.   TimTec

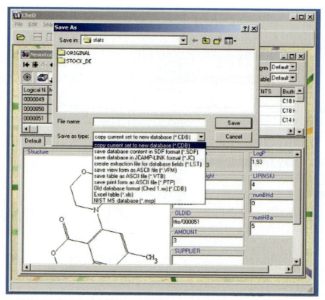

Figure 2.187.   TimTec

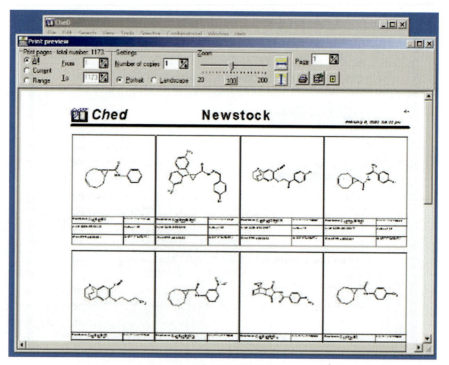

Figure 2.188.   TimTec

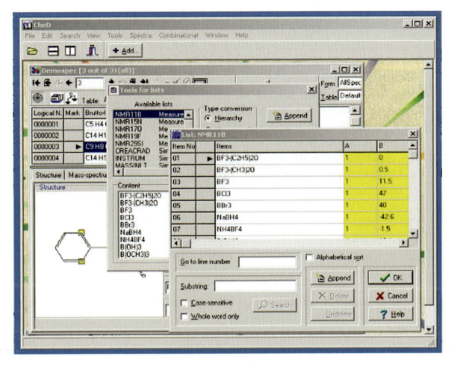

Figure 2.189.   TimTec

2.40 TRIPOS

I. Tripos, Inc.; http://www.tripos.com/

II. **Product Summaries:** *SYBYL/Base* is the core of Tripos Discovery Software for the dedicated molecular modeller. SYBYL/Base provides tools to build, edit, visualize and perform calculations on molecules of any size. The integrated Molecular Spreadsheet allows for the organization and analysis of molecular data. And the SYBYL Programming Language (SPL) is also provided to develop custom drug design methods. Tripos also offers a scientifically comprehensive set of modules for SYBYL/Base for Molecular modeling and visualization, Chemical informatics, Combinatorial chemistry and molecular diversity, Pharmacophore perception, Structure-activity relationship and ADME, Virtual Screening, Bioinformatics, and *de novo* design.

Benchware is Tripos' product line targeted at the laboratory scientist. Benchware applications provide data access, sharing and knowledge management capabilities through an n-tiered informatics platform built on open standard web based communication technologies. Benchware 3D Explorer provides an extensible platform for the visualisation, analysis and annotation and sharing of complex 3D chemical information. Benchware Dock is an underlying informatics platform that provides laboratory scientists with access to computational results and procedures developed by dedicated computational chemists. Benchware LibraryMaker and Library Designer provide parallel and combinatorial chemists desktop tools for enumerating and designing chemical libraries prior to synthesis. Benchware HTS DataMiner provides key scientific capabilities to rapidly mine HTS data for prospective lead series. Finally Benchware Notebook provides an electronic notebook capability to allow for ready protection of intellectual property data and capture and reuse of chemistry knowledge.

III. **Key capabilities and offerings:**

    a. **Computational Informatics products** – *Molecular modeling and visualization*.

        i. SYBYL/Base provides the general unified environment for computational molecular experiments of all types. SYBYL itself provides access to a wide array of molecular visualisation techniques, molecular mechanics calculations through a variety of force fields, data organisation through the molecular spreadsheet and access to the SPL programming and scripting language.

        ii. Advanced Computation provides a broad array of tools for thorough conformational analysis where the user can obtain all the possible torsional states of a molecule or only the low-energy conformations.

        iii. Confort performs rapid conformational searching and analyses of a drug-sized molecule as well as a specified substructure to identify the global minimum energy conformer, all local minima within a user-specified energy range, or a maximally diverse subset of conformers.

        iv. AMPAC can perform semi-empirical modeling calculations using several molecular wave functions (AM1, PM3, MINDO/3, MNDOC, MNDO/d, SAM1) for the determination of 3D structures and electronic properties of small molecules.

        v. HSCF is a unique semi-empirical molecular orbital (MO) "information server." HSCF was specifically designed for ease of use within other programs, scripts, and data pipeline protocols. It is ideally suited for use within QSPR applications in support of ADMET-related objectives.

        vi. MM4 force field is the latest in the well-known series of molecular mechanics programs developed by Prof. N. L. Allinger and co-workers at the University of Georgia. The high quality geometries and energies generated by these programs have led to their widespread recognition as the industry standard for calculating small molecule structures.

        vii. MOLCAD allows advanced visualization and manipulation capabilities for molecular surfaces, protein ribbon displays (tube, snake, cartoon, and secondary structure), slice and separated surfaces. Calculated molecular properties (e.g., lipophilic potential, Coulombic or Poisson-Boltzman electrostatic potential, hydrogen bonding) can be mapped onto any MOLCAD surface as color-coding.

viii. hint! provides tools to visualize and calculate the relative strengths of non-covalent interactions between and within biological molecules. These "hydropathic" interactions, which are the primary determinants of molecular recognition, include hydrogen-bonding, acid-base, Coulombic, and hydrophobic interactions

ix. ZAP uses the Poisson-Boltzmann equation to calculate the electrostatic potential surrounding a molecule in a medium of varying dielectric. The electrostatic potential contours calculated by ZAP can be simultaneously displayed with molecules for ease in interpretation. ZAP's electrostatic potential is readily used in the Comparative Molecular Field Analysis method found in QSAR with CoMFA. ZAP also calculates electrostatic descriptors that embody the surface properties of molecules and can be used in QSAR analyses for the creation of models to predict the properties of new molecules.

b. **Computational Informatics products – *Chemical Informatics*.**

i. AUSPYX is a data cartridge that allows the storage and searching of chemical structures and property data in Oracle8i relational databases.

ii. CONCORD is an engine that can work as a stand alone or with other computational software to rapidly convert 2-D to 3-D structures of compounds containing up to 200 heavy atoms. PowerCONCORD offers the same capabilities of CONCORD in addition to distributing and executing calculations in parallel for faster calculations.

iii. HiVol aids in the access and analysis of data generated from combinatorial chemistry, high-throughput synthesis and high-throughput screening activities. This tool can also calculate chemical descriptors including UNITY 2D Fingerprints, Clog P, CMR, number of rings, and rotatable bonds.

iv. ProtoPlex provides a mechanism for making accurate and realistic structural representations of chemical compounds. ProtoPlex allows users to control the protonation, deprotonation, or tautomerization for each chemical class of proto-centers, as well as the overall rules for formal charge and number of proto-centers to multiplex.

v. StereoPlex generates multiple stereoisomers of each input structure according to a user-specified rules. StereoPlex "multiplexes" both atom-centered (R/S) and bond-centered (E/Z) chirality.

vi. UNITY creates and searches 2-D and 3-D databases in order to identify structures that match a pharmacophore, fit a receptor site, or contain a desired chemical substructure. UNITY also provides tools to retrieve, manage, and analyze the resulting hits.

b. **Computational Informatics products – *Combinatorial chemistry and molecular diversity*.**

i. Legion/CombiLibMaker allows the construction and enumeration of virtual combinatorial libraries following user-specified filters. This module can use user-provided reactant lists to produce all possible compound permutations or sidechain groups to be applied to specified sites on a core structure. To minimize storage space and maximize data transfer of large virtual combinatorial libraries to other applications, Legion can save all the information in a 2D searchable combinatorial format (cSLN).

ii. DiverseSolutions calculates BCUT metrics for groups of molecules, provides a complete set of tools for chemical diversity analysis to identify and select compound subsets based on BCUTs and other metrics and compares the selected molecules to other sets of molecules. DiverseSolutions can be employed to generate both focussed and diverse libraries both for diversity selection and parallel and combinatorial library design

iii. OptDesign ids in the design of combinatorial libraries using practical constraints such as cost and ease of synthesis as well as user-defined filters (e.g., molecular weight, log P, excess reagent usage) to select a portion of a full library for synthesis. OptDesign can be employed to generate sparse as well as full matrix library designs of any user provided format.

iv. Selector expands the molecular descriptors included in Molecular Spreadsheet and provides a wide variety of analysis and selection algorithms to make diversity and similarity selections of compounds in the Molecular Spreadsheet or UNITY database.

c. **Computational Informatics products – *Pharmacophore perception.***

i. GALAHAD generates pharmacophore models for groups of active compounds and high quality molecular alignments. GALAHAD allows researchers to automatically develop pharmacophore hypotheses and structural alignments from a set of molecules that bind at a common site. No prior knowledge of pharmacophore elements, constraints, or molecular alignment is required.

ii. DISCOtech generates and refines multiple pharmacophore hypotheses from precomputed conformers of active target-based ligands. Developed phramacophores can be used for 3D database searching, ligand prediction and 3D QSAR studies

iii. FlexS superimposes ligands and aligns them for 3D QSAR studies and pharmacophore determinations. FlexS uses an incremental construction algorithm that grows the ligand onto the reference while optimizing the overlap of physico-chemical properties.

iv. GASP (Genetic Algorithm Similarity Program) elucidates pharmacophore models when or where no prior knowledge or ligand constraints are known.

v. RECEPTOR allows the analysis of a group of active ligands to elucidate the pharmacophore for the binding site of the target.

vi. Tuplets enables researchers to retrieve compounds from molecular structure databases that are pharmacophorically, and therefore biologically, similar to known actives without the need to derive a classical 3D pharmacophore model.

d. **Computational Informatics products –** *Structure-activity relationship and ADME.*

  i. QSAR with CoMFA provides a suite of tools to calculate molecular descriptors and develop statistical and graphical models of known biologically active compounds to predict the activity and properties of untested molecular candidates.

  ii. Advanced CoMFA expands the CoMFA fields found in the QSAR module allowing the user to refine predictive models.

  iii. ClogP/CMR calculates the hydrophobicity and molar refractivity values of chemical entities for QSAR applications.

  iv. Distill classifies molecular structures by substructures and displays the results to aid in the assessment of structure-activity relationships.

  v. HQSAR (Hologram QSAR) is a toolset designed for medicinal chemists that present a simplified user-friendly interface to analyze small and large compound sets for the generation of QSAR models and SAR profiles.

  vi. Molconn-Z calculates over 300 topological indices based on 2D and 3D molecular structures.

  vii. VolSurf calculates adsorption, distribution, metabolism, and excretion (ADME) models for small, medium, and large molecules, including DNA fragments, peptides, and proteins. Volsurf includes a number of validated models for important ADME properties such as intestinal absorption and blood brain barrier penetration.

  viii. Almond offers a unique set of 3D molecular descriptors and a statistical workbench with which to rapidly create and test structure-activity relationship hypotheses. Almond does not require the alignment of compounds prior to QSAR analysis.

  ix. GSSI provides an approach to modeling solution-phase properties from a consideration of solute-solvent interactions. GSSI can develop models for predicting various partition coefficients and membrane permeability coefficients in support of ADME-related efforts. GSSI can also be used to estimate the free energy of desolvation which accompanies ligand-receptor interaction.

  x. Savol rapidly calculates atomic and molecular surface area and volumes useful as descriptors for QSAR purposes.

e. **Computational Informatics products –** *Virtual Screening.*

  i. FlexX performs the fast docking of small conformationally flexible ligands to binding sites to virtually screen small compound libraries. FlexX can also carry out the incremental construction of flexible ligands in the binding site using molecular fragments.

  ii. CScore integrates multiple scoring functions to identify, evaluate and rank ligand-receptor binding modes and affinities.

  iii. CombiFlex allows the high-throughput docking of virtual compound libraries to identify what compounds are more likely to fit better at a

binding site aiding in the compound selection process for library design
and candidates synthesis.

iv. FlexX-Pharm uses receptor/ligand interaction and spatial location constraints as filters for docking speeding-up the calculation process and
producing enriched datasets.

v. FlexE adds the ability to consider protein structural variability while
docking with FlexX.

f. **Computational Informatics products – Bioinformatics.**

i. Biopolymer provides a toolset to create, fine-tune and visualize predictive models for biological molecules (e.g., proteins, DNA, RNA, and
carbohydrates).

ii. Composer allows the use of sequence and structural information from
known protein families for the 3D construction of unknown proteins.

iii. FUGUE analyzes the sequences of a target protein and compares them to
structural profiles of all known protein fold families.

iv. Genefold analyzes protein sequence data with those of known protein
structures and predicts 3D protein folding and function.

iv. MatchMaker takes amino acid sequences, compares the sequences with
known structures, and then predicts the 3D structure of the sequences.

v. ORCHESTRAR is a suite of applications designed for comparative protein modeling. With ORCHESTRAR you can create structural alignments of a set of homologs using sequence alignment information
involving a query sequence, use alignment information to calculate a
model of the query containing structurally conserved regions (SCRs),
search databases of solved structures to fill structurally variable regions
(SVRs) in an SCR model, and add sidechains using a rotamer library to
create a complete, all-atom homology model of a query sequence.

vi. ProTable (Protein Table) provides a toolset for the analysis, visualization,
and evaluation of protein structures.

vii. SiteID provides a selection of analysis and visualization tools to identify
potential binding sites for ligand-receptor or protein-protein interactions.

g. **Computational Informatics Products – *De novo* Design**

i. EA-Inventor is a completely generic *de novo* design engine or "toolkit"
which can be used in conjunction with any desired scoring function (or
composite scoring function).

ii. LeapFrog is a de novo ligand design tool that uses a target binding site or a
CoMFA model for the generation of new potentially active ligand structures.

iii. RACHEL is a tool for lead compound optimization and ligand design where
a given scaffold is systematically combinatorialized and evaluated within a
binding receptor to provide novel and enriched high-affinity ligands.

h. **Laboratory Informatics Products**

i. Benchware 3D Explorer provides access to 3D structural data visualisation, analysis, knowledge capture and sharing. Very high quality visualisation combined with contextual structure modification and extensive

annotation and captioning capabilities aid laboratory scientsts in understanding structural requirements for biological properties and enable
what if experiments. 3D Explorer can be extensively modified through
VBA and integrated with in-place informatics systems through
published APIs.

ii. Benchware Dock allows molecular modellers to publish the results of
their work into a shared database system. Modeling results can be
accessed by chemists to allow them to view predicted structures and test
new synthesis candidates for compatibility with a target receptor structure prior to synthesis. Benchware Dock is built upon the Tripos foundation server and ModelBase technologies.

iii. Benchware HTS DataMiner is an application for the processing and analysis of large biological and chemical datasets and libraries to determine candidate lead series from HTS data, develop SAR models and expand
structure selection beyond the original HTS series. Benchware HTS
Dataminer removes the analysis bottleneck in the hits to leads process.

iv. Benchware Librarymaker allows laboratory scientists to rapidly enumerate parallel and combinatorial libraries rapidly generating electroinic
copies of product structure molecules

v. Benchware LibraryDesigner allows laboratory scientists access to
validated library design techniques that utilise the same technology that
is available through the DiverseSolutions product. Library designer
allows chemists to balance the desired physicochemical properties of the
libraries they wish to synthesise with the economy of actualy synthesis-
ing those libaries. By allowing chemists to utilise library design technology directly rather than relying on input from computational groups
design cycles and synthesis cycles can be completed much more rapidly.

vi. Benchware Notebook is a full features electronic notebook for chemistry
operations. The notebook is built on the same Tripos platform technology
components as used in Benchware Dock and has been designed with ease of
use and integration in mind. Organisations using Benchware notebook benefit from IP protection, enhanced productivity through page template use
and cloning and most importantly capture and reuse of synthetic knowledge.

i. **Platform technology products**

i. Tripos Foundation Server offers access to Tripos and 3rd party applications as web services. Utilising the Tripos foundation Server allows
organisations the ability to deploy scientific compute engines and
protocols derived from them to a wide variety of researchers and client
interfaces within their organisaion.

ii. ModelBase is an Oracle based database system for the storage and sharing of computational models. ModelBase allows molecular modellers to
readily share and build upon each others models, avoiding duplication of
effort and loss of knowledge during personnel changes. ModelBase is
employed as the basis for the Benchware Dock product.

IV. **Review**:

Tripos Inc. provides a comprehensive software suite and modules to aid scientists in different research fields in drug discovery. The SYBYL®/Base platform integrates all the available modules running under the same environment and user can access them using the main interface which can be customized, and it also allows the user to create new computational methods via the SYBYL Programming Language (SPL). The SYBYL®/Base platform and its modules are available for the IRIX 6.5, HP-UX 11i, and Linux operating systems. Tripos' Alchemy 2000™; and SARNavigator™; run under Microsoft Windows 95 or higher, so Windows users will find a familiar interface and easy to navigate, identify and use the tools in these 2 programs. Currently, SYBYL®/Base requires knowledge of UNIX to install it and maintain it, a fact that limits the user audience. It takes time to become familiarized with the SYBYL®/Base graphical user interface (GUI) as many of the icons displayed are not very descriptive. The great flexibility and realm of options found in SYBYL®/Base and its modules ensures the user has at her/his disposal what is needed to get things done; however, this could lead to the overlook of use/selection of necessary parameters needed for some tasks resulting in incorrect results for which little-to-no description is offert to pinpoint the problem. Nevertheless, once the user is familiarized with the GUI and processes of interest, SYBYL provides a robust platform for drug discovery.

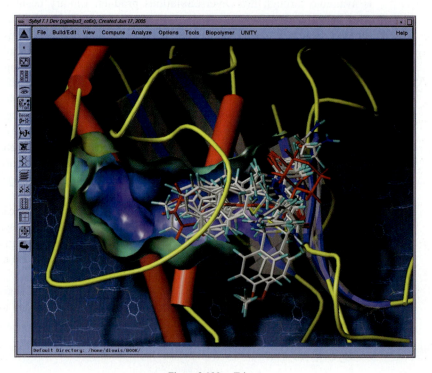

Figure 2.190.   Tripos

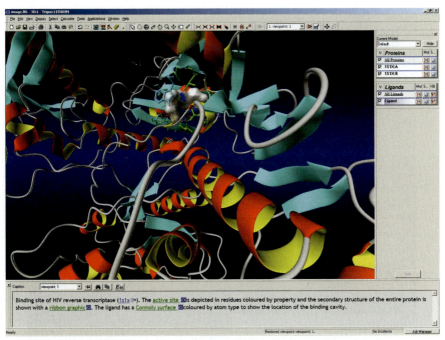

Figure 2.191. Tripos

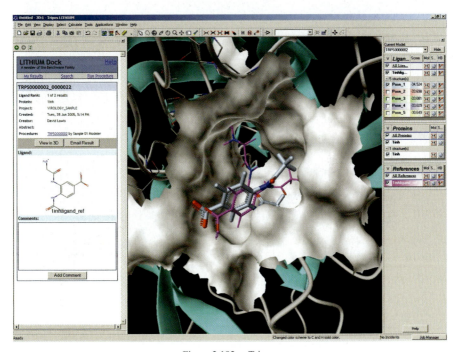

Figure 2.192. Tripos

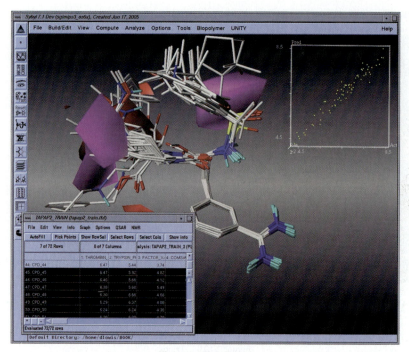

Figure 2.193.   Tripos

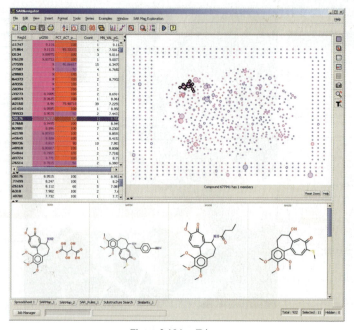

Figure 2.194.   Tripos

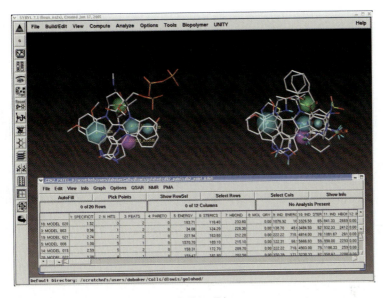

Figure 2.195.   Tripos

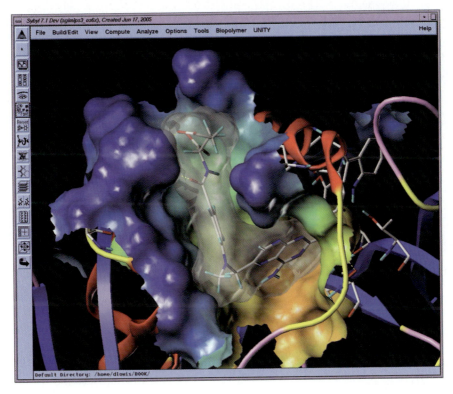

Figure 2.196.   Tripos

SUBJECT APPENDICES

DRUG DISCOVERY INFORMATICS REGISTRATION SYSTEMS
AND UNDERLYING TOOLKITS (APPENDICES 1 AND 2)

Appendix 1:   DRUG, MOLECULAR REGISTRATION SYSTEMS, AND CHEMISTRY DATA
CARTRIDGES

Product	Company	Page	Website
Isentris, MDL-Base, ISIS-Base	MDL	171	http://www.mdli.com
ChemOffice	Cambridgesoft	73	http://www.cambridgesoft.com
Cambridgesoft Enterprise Solutions	CambridgeSoft	73	http://www.cambridgesoft.com/solutions/
ActivityBase	IDBS	152	http://www.idbs.com/products/abase/
Accord	Accelrys	51	http://www.accelrys.com
AUSPYX	Tripos	261	http://www.tripos.com
CBIS (Chemical and Biological Informatics Systems)	ChemInnovation Software	103	http://www.cheminnovation.com/
ACD/SpecDB (Spectral Data Management System) - Analytical data capture tools	ACD Labs	59	http://www.acdlabs.com
Super Package /THOR	BioByte (Daylight)	71	http://www.biobyte.com/index.html
JChem Cartridge	ChemAxon	89	http://www.chemaxon.com
Software for Chemical databases and chemical data processes	Infochem	156	http://www.infochem.de/en/company/index.shtml

The InfoChem Chemistry Cartridge for Oracle	Infochem	156	http://www.infochem.de/eng/index.htm
Molecular Databank	Collaborative Drug Discovery	115	http://www.collaborativedrug.com/
DayCart	Daylight Chemical Information Systems, Inc.	123	http://www.daylight.com/
CHED, Chemdbsoft	TimTec	254	http://software.timtec.net/
MDL Isentris platform (ISIS Base), ChemBio AE, AssayExplorer	MDL	171	http://www.mdli.com

Appendix 2:   CHEMOINFORMATICS TOOLKITS TO DEVELOP APPLICATIONS

Product	Company	Page	Website
OEChem	Open Eye	194	http://www.eyesopen.com/
Marvin	ChemAxon	87	http://www.chemaxon.com/
Calculator Plugins	ChemAxon	88	http://www.chemaxon.com
JChem Base	ChemAxon	89	http://www.chemaxon.com
MOE (Molecular Operating Environment)	Chemical Computing Group	98	http://www.chemcomp.com/
CORINA + other products	Molecular Networks GmbH	187	http://www.mol-net.com
MetaDrug	GeneGo	141	http://www.genego.com

CONTENT DATABASES (APPENDICES 3–7)

Appendix 3:   COMPOUND AVAILABILITY DATABASES

Product	Company	Page	Website
Available Chemicals Directory (ACD, ACD screen)	MDL	171	http://www.mdli.com
ChemACX	Cambridgesoft	73	http://www.cambridgesoft.com/databases/
ChemMatrix Technology	ChemNavigator	109	http://www.chemnavigator.com
PubChem	PubChem	203	http://pubchem.ncbi.nlm.nih.gov/
Molecular Databank	Collaborative Drug Discovery	115	http://www.collaborativedrug.com/

Appendix 4:   SAR DATABASE

Product	Company	Page	Website
Kinase ChemBioBase	Jubilant Biosys	164	http://www.jubilantbiosys.com/
Gene family databases	GVK-BIO	144	http://www.gvkbio.com/
LUCIA	Eidogen-Sertanty	128	http://www.eidogen-sertanty.com/
TIP (Target Informatics Platform)	Eidogen-Sertanty	127	http://www.eidogen-sertanty.com/
GPCR Annotator	Jubilant Biosys	164	http://www.jubilantbiosys.com/
PubChem	PubChem	203	http://pubchem.ncbi.nlm.nih.gov/
CDD Molecular Databank	Collaborative Drug Discovery	115	http://www.collaborativedrug.com/
MDDR (MDL Drug Data Report)	MDL	175	http://www.mdli.com

Appendix 5:   CHEMICAL REACTION DATABASES

Product	Company	Page	Website
SPREZZI	InfoChem	156	http://www.infochem.de/en/company/index.shtml
SPRESI^{web}	InfoChem	158	http://www.infochem.de/en/company/index.shtml
Beilstein, Crossfire, SPORE, ChemInform Reaction Library, ORGSYN, etc.	MDL	175	http://www.mdli.com
LUCIA	Sertanty (previously Libraria)	128	http://www.eidogen-sertanty.com
Arthur	Symyx	248	http://www.symyx.com
SciFinder	Chemistry Abstract Services	80	http://www.cas.org, http://www.cas.org/prod.html
CA File	Chemistry Abstract Services	80	http://www.cas.org, http://www.cas.org/prod.html
CAS Registry	Chemistry Abstract Services	80	http://www.cas.org, http://www.cas.org/prod.html
CA on CD	Chemistry Abstract Services	81	http://www.cas.org, http://www.cas.org/prod.html
CAplus	Chemistry Abstract Services	81	http://www.cas.org, http://www.cas.org/prod.html
MARPAT – CAS Markush database	Chemistry Abstract Services	81	http://www.cas.org, http://www.cas.org/prod.html
CASREACT	Chemistry Abstract Services	81	http://www.cas.org, http://www.cas.org/prod.html
CHEMCATS File	Chemistry Abstract Services	82	http://www.cas.org, http://www.cas.org/prod.html
RxD – Chemical Reaction Database	GVK BIO	144	http://www.gvkbio.com/

Appendix 6:   PATENT DATABASES

Product	Company	Page	Website
Derwent	Thompson /MDL	175	http://www.mdli.com
MDL Patent Chemistry Database	MDL	176	http://www.mdli.com
CAplus	Chemistry Abstract Services	81	http://www.cas.org, http://www.cas.org/prod.html
MARPAT – CAS Markush database	Chemistry Abstract Services	81	http://www.cas.org, http://www.cas.org/prod.html

Appendix 7:   OTHER COMPOUND AND DRUG DATABASES

Product	Company	Page	Website
MDDR (MDL Drug Data Report), DiscoveryGate, OHS Reference, xPharm, etc.	Prous/MDL	175	http://www.mdli.com
MetaCore	GeneGo	140	http://www.genego.com
MCD – Medichem Database	GVK BIO	144	http://www.gvkbio.com/
TID – Target inhibitor databases	GVK BIO	145	http://www.gvkbio.com/
PKD – Pharmacokinetic	GVK BIO	145	http://www.gvkbio.com/
PPID – Protein-Protein interactions	GVK BIO	145	http://www.gvkbio.com/
PathArt©	Jubilant Biosys	164	http://www.jubilantbiosys.com/
Nitrilase and Nitrile Hydratase Knowledgebase	Jubilant Biosys	165	http://www.jubilantbiosys.com/
Cambridgesoft databases	CambridgeSoft	75	http://www.cambridgesoft.com/databases/
Global Major Reference Works	InfoChem (and MDL)	158	http://www.infochem.de/en/company/index.shtml

DRUG, MOLECULE, AND PROTEIN VISUALIZATION (APPENDICES 8–10)

Appendix 8:   CHEMICAL DRAWING, STRUCTURE VIEWING AND MODELING PACKAGES

Product	Company	Page	Website
Chemdraw (8.0)	CambridgeSoft	73	http://www.cambridgesoft.com
ChemOffice	CambridgeSoft	73	http://www.cambridgesoft.com
HyperChem Web Viewer	HyperCube	149	http://www.hyper.com/default.htm
Chemlab	Scinova Technologies	224	http://www.scinovaindia.com/
ArgusLab	Planaria-Software	202	http://www.planaria-software.com/
RasMol – Protein Explorer	Academic (non-commercial)	211	http://www.umass.edu/microbio/rasmol/
Marvin	ChemAxon	87	http://www.chemaxon.com
Chem-4D	ChemInnovation Software	103	http://www.cheminnovation.com/
Chemsite	ChemInnovation Software	104	http://www.cheminnovation.com/
Cerius^2 Discovery Studio ViewerPro	Accelrys	51	http://www.accelrys.com
PyMol 0.92	PyMol	208	http://www.pymol.org
Chemical Database Management Software	TimTec	254	http://software.timtec.net/
ISIS-Draw (MDL Draw)	MDL	173	http://www.mdli.com
Chime plug-in (Ras-Mol)	MDL	211	http://www.mdli.com
Sybyl-base	Tripos	259	http://www.tripos.com/

Appendix 9:   DATA ANALYSIS AND MINING TOOLS

Product	Company	Page	Website
Pipeline Pilot	SciTegic	226	http://www.scitegic.com/products_ services/pipeline_pilot.htm
Prometheus	Scinova Technologies	224	http://www.scinovaindia.com/
Discovery Studio	Accelrys	52	http://www.accelrys.com
CombiMat 2.0	Accelrys	51	http://www.accelrys.com
Screen	Chemaxon	90	http://www.chemaxon.com/

JKlustor	Chemaxon	90	http://www.chemaxon.com/
Reactor	Chemaxon	90	http://www.chemaxon.com/
Fragmentor	Chemaxon	90	http://www.chemaxon.com/
Insight II	Accelrys	51	http://www.accelrys.com
STS (Synthesis Tree Search)	Infochem	158	http://www.infochem.de/en/company/index.shtml
CLASSIFY	Infochem	157	http://www.infochem.de/en/company/index.shtml
IC-Map	Infochem	157	http://www.infochem.de/en/company/index.shtml
The ICFSE (Infochem Fast Search Engine)	Infochem	157	http://www.infochem.de/en/company index.shtml
Filter, Omega, ROCS, SMACK, etc.	Open Eye	194	http://www.eyesopen.com/
DiverseSolutions	Tripos	262	http://www.tripos.com/
Spotfire DecisionSite	Spotfire	236	http://www.spotfire.com
Diversity Analysis, Cluster Analysis	Barnard Chemical Information Ltd. (BCI);	68	http://www.bci.gb.com/
CAChe, BioMedCAChe	Fujitsu Biosciences Group	137	http://us.fujitsu.com/biosciences
Merlin *(exploratory data analysis tools)*	Daylight Chemical Information Systems, Inc.	123	http://www.daylight.com/
Thor *(a database system based on Daylight's SMILES notation)*	Daylight Chemical Information Systems, Inc.	123	http://www.daylight.com/

Appendix 10:   SMALL MOLECULE – PROTEIN VISUALIZATION TOOLS

Product	Company	Page	Website
RasMol	University of Massachusetts	211	http://www.umass.edu/micro bio/rasmol/
Protein Explorer	US National Science Foundation	211	http://molvis.sdsc.edu/protex pl/frntdoor.htm
3-Dimensional Protein-Ligand Map (3-DPL)	ChemNavigator	109	http://www.chemnavigator.com
Eve (Eidogen Visualization Environment)TM	Eidogen-Sertanty	128	http://www.eidogen-sertanty.com/

PyMOL	PyMOL	208	http://www.pymol.org; hosted at: http://pymol.sourceforge.net/
3-D molecular visualization program	Chimera	112	http://www.cgl.ucsf.edu/chimera/index.html
Molsmart – converts MDL to Daylight formats	Barnard Chemical Information Ltd. (BCI);	68	http://www.bci.gb.com/
MOE (Molecular Operating Environment)	Chemical Computing Group	98	http://www.chemcomp.com/
MDL Sculpt	MDL	174	http://www.mdli.com

MODELING AND ALGORITHMS (APPENDICES 11–17)

Appendix 11:   MOLECULAR DESCRIPTORS

Product	Company	Page	Website
DRAGON	Milano Chemometrics and QSAR Research Group	180	http://www.disat.unimib.it/chm/
MOBYDIGS	Milano Chemometrics and QSAR Research Group	181	http://www.disat.unimib.it/chm/
KOALA	Milano Chemometrics and QSAR Research Group	181	http://www.disat.unimib.it/chm/
DOLPHIN	Milano Chemometrics and QSAR Research Group	181	http://www.disat.unimib.it/chm/
Chemical Structure Fragments and Fingerprinting.	Barnard Chemical Information Ltd. (BCI)	68	http://www.bci.gb.com/
Calculator Pluggins	Chemaxon	88	http://www.chemaxon.com/
Daylight fingerprints	Daylight Chemical Information Systems, Inc.	123	http://www.daylight.com/
OEChem, Ogham, and related Apps	Open Eye	194	http://www.eyesopen.com/
AMPAC, Confort, MM3(2000), MOLCAD, and related Apps	Tripos	260	http://www.tripos.com/

Appendix 12:   CLOGP, TPSA, AND LIPINSKI PROPERTY CALCULATION SYSTEMS

Product	Company	Page	Website
Regular Package/ MASTERFILE	BioByte (Daylight)	71	http://www.biobyte.com/index.html
MERLIN	BioByte	71	http://www.biobyte.com/index.html
BDRIVE	BioByte	71	http://www.biobyte.com/index.html
Economy Package/ UDRIVE	BioByte	71	http://www.biobyte.com/index.html
Calculator Pluggins	Chemaxon	88	http://www.chemaxon.com/
Compound Property Predictors (for physical chemical properties and spectra)	Advanced Chemistry Development (ACD Labs)	60	http://www.acdlabs.com
PASS, Slipper, etc.	TimTec	254	http://software.timtec.net/

Appendix 13:   QSAR/PHARMACOPHORE PROGRAMS

Product	Company	Page	Website
Catalyst	Accelrys	51	http://www.accelrys.com
Materials Studio	Accelrys	51	http://www.accelrys.com
Catalyst	Accelrys	51	http://www.accelrys.com
QSAR with CoMFA	Tripos	263	http://www.tripos.com/
DISCOtech	Tripos	262	http://www.tripos.com/
GASP	Tripos	262	http://www.tripos.com/
RECEPTOR	Tripos	262	http://www.tripos.com/
C-QSAR	Biobyte	69	http://www.biobyte.com/index.html
BIO/PHYS	Biobyte	69	http://www.biobyte.com/index.html
SIGMA – A THOR	Biobyte	70	http://www.biobyte.com/index.html
Chemical Structure Fragments and Fingerprinting.	Barnard Chemical Information Ltd. (BCI)	68	http://www.bci.gb.com/
MOE (Molecular Operating Environment)	Chemical Computing Group	98	http://www.chemcomp.com/
MDL QSAR	MDL	174	http://www.mdli.com

Appendix 14: DOCKING AND CRYSTALLOGRAPHIC SOFTWARE

Product	Company	Page	Website
Quanta	Accelrys	52	http://www.accelrys.com
FirstDiscovery, Glide	Schrödinger	215	http://www.schrodinger.com/
ArgusLab	Planaria-Software	202	http://www.planaria-software.com/
CombiFlex, CScore, FlexS, and FlexX	Tripos	263	http://www.tripos.com
FRED	Open Eye	196	http://www.eyesopen.com/
DOCK	UCSF	112	http://dock.compbio.ucsf.edu/index.html
View Dock	UCSF	112	http://www.cgl.ucsf.edu/chimera/docs/ ContributedSoftware/viewdock/ viewdock.html

Appendix 15: QUANTUM MECHANICS CALCULATIONS

Product	Company	Page	Website
MacroModel®	Schrödinger	216	http://www.schrodinger.com/
Jaguar, *Qsite*	Schrödinger	215	http://www.schrodinger.com/
Density Functional Theory/ HyperChem 7	Hypercube	148	http://www.hyper.com/default.htm
HyperChem 7.5	HyperCube	148	http://www.hyper.com/default.htm
HyperChem Lite 2.0	HyperCube	149	http://www.hyper.com/default.htm
MOPAC 2000, Quantum CAChe	Fujitsu Biosciences Group	138	http://us.fujitsu.com/biosciences

Appendix 16: PK/ADME/TOX DATABASES AND PREDICTORS

Product	Company	Page	Website
PK Solutions (PK)	Summit PK	239	http://www.summitpk.com/
Metabase (PK/ADME/Tox)	Summit PK	239	http://www.summitpk.com/
MetaDrug	GeneGo	141	http://www.genego.com
ToxScope	LeadScope	170	http://www.leadscope.com/
PENGUINS	Molecular Discovery	185	http://www.moldiscovery.com/index.php

QikPro	Schrödinger	216	http://www.schrodinger.com/
MDL Metabolite Database, Toxicity Database	MDL	175	http://www.mdli.com
GastroPlus	Simulation Plus, Inc.	229	http://www.simulations-plus.com/
QMPRchitect,	Simulation Plus, Inc.	229	http://www.simulations-plus.com/

Appendix 17: MULTI-PARAMETER DRUG DEVELOPMENT/IDENTIFICATION SOFTWARE

Product	Company	Page	Website
LeadScope Enterprise	LeadScope	169	http://www.leadscope.com/
Personal LeadScope	LeadScope	169	http://www.leadscope.com/
ChemMatrix Technology	ChemNavigator	109	http://www.chemnavigator.com
GRID	Molecular Discovery	184	http://www.moldiscovery.com/index.php
VolSURF	Molecular Discovery	185	http://www.moldiscovery.com/index.php
Almond	Molecular Discovery	185	http://www.moldiscovery.com/index.php
MetaSite	Molecular Discovery	185	http://www.moldiscovery.com/index.php
ChIP™	Eidogen-Sertanty	128	http://www.eidogen-sertanty.com/
Rx	Scinova Technologies	224	http://www.scinovaindia.com/
QMPRPlus	Simulation Plus, Inc.	229	http://www.simulations-plus.com/
Spotfire DecisionSite for Lead Discovery	Spotfire	236	http://www.spotfire.com

INDEX